# TECHNOLOGICAL OBSOLESCENCE MANAGEMENT OF NUCLEAR POWER PLANTS

The following States are Members of the International Atomic Energy Agency:

AFGHANISTAN
ALBANIA
ALGERIA
ANGOLA
ANTIGUA AND BARBUDA
ARGENTINA
ARMENIA
AUSTRALIA
AUSTRIA
AZERBAIJAN
BAHAMAS, THE
BAHRAIN
BANGLADESH
BARBADOS
BELARUS
BELGIUM
BELIZE
BENIN
BOLIVIA, PLURINATIONAL
  STATE OF
BOSNIA AND HERZEGOVINA
BOTSWANA
BRAZIL
BRUNEI DARUSSALAM
BULGARIA
BURKINA FASO
BURUNDI
CABO VERDE
CAMBODIA
CAMEROON
CANADA
CENTRAL AFRICAN
  REPUBLIC
CHAD
CHILE
CHINA
COLOMBIA
COMOROS
CONGO
COOK ISLANDS
COSTA RICA
CÔTE D'IVOIRE
CROATIA
CUBA
CYPRUS
CZECH REPUBLIC
DEMOCRATIC REPUBLIC
  OF THE CONGO
DENMARK
DJIBOUTI
DOMINICA
DOMINICAN REPUBLIC
ECUADOR
EGYPT
EL SALVADOR
ERITREA
ESTONIA
ESWATINI
ETHIOPIA
FIJI
FINLAND
FRANCE
GABON
GAMBIA, THE

GEORGIA
GERMANY
GHANA
GREECE
GRENADA
GUATEMALA
GUINEA
GUYANA
HAITI
HOLY SEE
HONDURAS
HUNGARY
ICELAND
INDIA
INDONESIA
IRAN, ISLAMIC REPUBLIC OF
IRAQ
IRELAND
ISRAEL
ITALY
JAMAICA
JAPAN
JORDAN
KAZAKHSTAN
KENYA
KOREA, REPUBLIC OF
KUWAIT
KYRGYZSTAN
LAO PEOPLE'S DEMOCRATIC
  REPUBLIC
LATVIA
LEBANON
LESOTHO
LIBERIA
LIBYA
LIECHTENSTEIN
LITHUANIA
LUXEMBOURG
MADAGASCAR
MALAWI
MALAYSIA
MALI
MALTA
MARSHALL ISLANDS
MAURITANIA
MAURITIUS
MEXICO
MONACO
MONGOLIA
MONTENEGRO
MOROCCO
MOZAMBIQUE
MYANMAR
NAMIBIA
NEPAL
NETHERLANDS,
  KINGDOM OF THE
NEW ZEALAND
NICARAGUA
NIGER
NIGERIA
NORTH MACEDONIA
NORWAY
OMAN

PAKISTAN
PALAU
PANAMA
PAPUA NEW GUINEA
PARAGUAY
PERU
PHILIPPINES
POLAND
PORTUGAL
QATAR
REPUBLIC OF MOLDOVA
ROMANIA
RUSSIAN FEDERATION
RWANDA
SAINT KITTS AND NEVIS
SAINT LUCIA
SAINT VINCENT AND
  THE GRENADINES
SAMOA
SAN MARINO
SAUDI ARABIA
SENEGAL
SERBIA
SEYCHELLES
SIERRA LEONE
SINGAPORE
SLOVAKIA
SLOVENIA
SOMALIA
SOUTH AFRICA
SPAIN
SRI LANKA
SUDAN
SWEDEN
SWITZERLAND
SYRIAN ARAB REPUBLIC
TAJIKISTAN
THAILAND
TOGO
TONGA
TRINIDAD AND TOBAGO
TUNISIA
TÜRKİYE
TURKMENISTAN
UGANDA
UKRAINE
UNITED ARAB EMIRATES
UNITED KINGDOM OF
  GREAT BRITAIN AND
  NORTHERN IRELAND
UNITED REPUBLIC OF TANZANIA
UNITED STATES OF AMERICA
URUGUAY
UZBEKISTAN
VANUATU
VENEZUELA, BOLIVARIAN
  REPUBLIC OF
VIET NAM
YEMEN
ZAMBIA
ZIMBABWE

The Agency's Statute was approved on 23 October 1956 by the Conference on the Statute of the IAEA held at United Nations Headquarters, New York; it entered into force on 29 July 1957. The Headquarters of the Agency are situated in Vienna. Its principal objective is "to accelerate and enlarge the contribution of atomic energy to peace, health and prosperity throughout the world".

IAEA-TECDOC-2112

# TECHNOLOGICAL OBSOLESCENCE MANAGEMENT OF NUCLEAR POWER PLANTS

INTERNATIONAL ATOMIC ENERGY AGENCY
VIENNA, 2026

For further information on this publication, please contact:

Operational Safety Section
International Atomic Energy Agency
Vienna International Centre
PO Box 100
1400 Vienna, Austria
Email: Official.Mail@iaea.org

© IAEA, 2026
Printed by the IAEA in Austria
January 2026
https://doi.org/10.61092/iaea.1z3q-gtl7

**IAEA Library Cataloguing in Publication Data**

Names: International Atomic Energy Agency.
Title: Technological obsolescence management of nuclear power plants / International Atomic Energy Agency.
Description: Vienna : International Atomic Energy Agency, 2026. | Series: IAEA-TECDOC ISSN 1011-4289 ; no. 2112 | Includes bibliographical references.
Identifiers: IAEAL 26-01810 | ISBN 978-92-0-129325-1 (paperback : alk. paper) | ISBN 978-92-0-129425-8 (pdf)
Subjects: LCSH: Nuclear power plants — Safety measures. | Nuclear power plants — Management. | Depreciation.

# FOREWORD

To assist its Member States in managing the ageing of nuclear power plants, the IAEA launched the International Generic Ageing Lessons Learned (IGALL) programme in 2009. IGALL provides a collection of proven practices for implementation of ageing management, based on examples shared by Member States. In Phase 2 of the IGALL programme, it was recognized by Member States that technological obsolescence is an area where relevant national practices could be collected and shared. The IGALL Steering Committee, consisting of representatives of 30 IAEA Member States operating nuclear power plants, therefore tasked the IGALL programme with developing a technological obsolescence management framework. In response, the Technological Obsolescence Programme was developed, which follows a format similar to that of the IGALL ageing management programmes. The Technological Obsolescence Programme documentation describes the nine attributes of an effective obsolescence management programme.

Although the Technological Obsolescence Programme defines the attributes of an effective obsolescence management programme, it does not provide implementation lessons learned from Member States or information on how to develop adequate training programmes related to obsolescence management. This publication aims to address that gap by presenting examples of implementation from Member States.

The IAEA is grateful for the contributions of all those who were involved in the drafting and review of this publication. The IAEA officers responsible for this publication were M. Marchena and B. Lehman of the Division of Nuclear Installation Safety.

# CONTENTS

# 1. INTRODUCTION

## 1.1. BACKGROUND

More than two-thirds of the world's operational nuclear power plants (NPPs) have been in operation for more than 30 years and over half have been operating for more than 40 years. There is growing recognition that continuing to operate these plants beyond their planned lifetimes can have economic and environmental benefits, but it also raises concerns about their long term safety and reliability. As a result, there is an increasing need for management strategies to ensure safe long term operation (LTO).

The ageing of NPPs includes both physical and non-physical effects. Physical ageing includes degradation phenomena such as corrosion, fatigue, and thermal ageing, among other factors. Non-physical ageing includes the obsolescence of structures, systems and components (SSCs). Obsolescence can occur due to components becoming unavailable in the market, suppliers no longer being available, changes in technology, or unavailability of spare parts. These aspects are included in so-called 'technological obsolescence'. 'Conceptual obsolescence' can occur due to changes in regulations, codes, and standards, or if related knowledge becomes obsolete. This document focuses on 'technological obsolescence' (hereafter referred to as 'obsolescence'). Both types of ageing, physical and non-physical ageing, require careful management to ensure the continued safe and efficient operation of the plant.

To address these challenges, plant operating organizations have developed comprehensive strategies that include ageing management programmes (AMPs), time-limited ageing analysis, and maintenance, surveillance and inspection activities. Additionally, they have plans in place to address obsolescence of SSCs, which may require the identification of obsolescence issues, prioritization of obsolete items, and implementation of solutions such as alternative SSCs, replacement of individual components, and in some cases the replacement of entire systems. While in some cases obsolescence is managed in a reactive manner, it is desirable to have a proactive programme that can anticipate unavailability of equipment.

Operating organizations develop their programmes to address their specific needs based on available tools, the regulatory framework, and their specific relationships with other operating organizations and suppliers. The purpose of this TECDOC is to provide operating organizations with a collection of practices and experiences related to the development and implementation of a proactive obsolescence management programme, and its impact on the availability of SSCs.

To assist States in managing ageing effectively, the IAEA developed a programme on International Generic Ageing Lessons Learned (IGALL). IGALL is a collection of proven practices for the implementation of ageing management based on input provided by Member States. During Phase 5 (2020–2021) of the IGALL programme, the IGALL Steering Committee requested the IAEA to collect operating experience and proven practices for managing technological obsolescence. Several Member States provided operating experience, which was used as the basis for this TECDOC.

Even though a technological obsolescence programme is not strictly an AMP, the proven practices and lessons learned described in this TECDOC consider the nine attributes of an effective AMP based on IAEA Safety Standards Series No. SSG-48, Ageing Management and Development of a Programme for Long Term Operation of Nuclear Power Plants [1], and it is also aligned with IAEA Safety Standards Series No. SSR-2/2 (Rev. 1), Safety of Nuclear Power Plants: Commissioning and Operation [2].

This publication complements:

— IAEA Nuclear Energy Series Nos: NR-T-3.34, Management of Ageing and Obsolescence of Instrumentation and Control Systems and Equipment in Nuclear Power Plants and Related Facilities Through Modernization [3]; and NP-T-3.21, Procurement Engineering and Supply Chain Guidelines in Support of Operation and Maintenance of Nuclear Facilities [4].

— IAEA Safety Reports Series No. 82, Ageing Management for Nuclear Power Plants: International Generic Ageing Lessons Learned (IGALL) [5].

— TOP401, Technological Obsolescence Programme [6], which was developed within the framework of the IGALL programme and is fully aligned with Ref. [5]. The TOP401 document can be found on the publicly available IAEA IGALL Programme website.

This publication summarizes the information collected from 10 Member States that indicated their willingness to share their experience. This information was collected during Phase 5 and 6 of the IGALL programme between 2020 and 2024, and during consultancy meetings in 2023 and 2024.

## 1.2. OBJECTIVE

The objective of this publication is to provide practical information on Member States' approaches and practices related to the development and implementation of proactive technological obsolescence programmes, including lessons learned. This publication also considers the regulatory framework and oversight of the obsolescence process in Member States.

This TECDOC is intended for use by operating organizations, regulatory bodies, supply chain organizations (e.g. manufacturers, designers) and technical support organizations.

## 1.3. SCOPE

The scope of the TECDOC covers lessons learned from Member States' implementation of proactive technological obsolescence management programmes. It covers SSCs important to safety and spare parts required to maintain those SSCs and may be applied to all SSCs important for plant reliability and availability. This publication focuses on experience from NPPs; however, it also includes experience from a fuel cycle facility and the information contained in this publication may be useful to other nuclear installations.

## 1.4. STRUCTURE

This publication summarizes the information provided by the Member States and is structured in six main sections: Section 2 summarizes the experience of NPPs in the development of obsolescence management programmes; Section 3 provides a general description of a proactive obsolescence management programme based on key components of a programme; Section 4 describes the involvement of the regulatory body in obsolescence management; Section 5 describes Member States' experiences with training of personnel on obsolescence management; Section 6 describes operating experience and real examples of obsolescence solutions.

The Annex presents additional details on the experiences of Member States in developing and implementing obsolescence management programmes, including the complete information that was provided by the Member States.

## 2. EXPERIENCE IN THE DEVELOPMENT OF OBSOLESCENCE MANAGEMENT PROGRAMMES

This section provides general expectations for obsolescence management and highlights Member State's experience with implementation of the programmes.

For plants that are at the initial stages of developing an obsolescence management programme, external support or reviews by industry organizations such as the International Nuclear Utility Obsolescence Group (INUOG), Nuclear Utility Obsolescence Group (NUOG), Electric Power Research Institute (EPRI), and peer reviews have proven to be an effective and useful tool to exchange experience, lessons learned and improve implementation. Sharing of knowledge, engagement with other organizations, and peer to peer connections save a lot of time and are beneficial for setting up a successful obsolescence management programme. Support from industry organizations can be leveraged from the initial steps of development through the whole management of the programme for continuous improvement.

The IAEA Safety Aspects of Long Term Operation (SALTO) missions have helped make the subject of obsolescence management more visible to management and senior management of NPPs and to regulatory bodies. These ageing management and LTO related reviews identify whether obsolescence management aligns with the IAEA Safety Standards. At the same time, such reviews may help identify the need to improve proactive obsolescence management. These missions have also provided the operating organizations of NPPs with an opportunity to exchange experiences and learn what was needed for the development of an obsolescence management programme.

Prior to implementing a formal obsolescence management programme, NPPs are also encouraged to first evaluate their existing processes, identify which aspects are relevant to managing obsolescence, and then identify the gaps within the process. This will help set the basis for customizing an obsolescence management programme that fits the NPP. Additionally, it is encouraged to implement measures that enable responsible personnel to recognize and report any issues related to obsolescence. Ensuring proper prioritization and mitigation of these issues can be achieved through established change management processes or using specialized tools. Some NPPs have achieved this through a tracking system that can be used throughout the plant to report identification of obsolescence issues. Other NPPs use internal intranet as a common location for documenting items.

Developing an obsolescence management programme requires the involvement of experienced and knowledgeable staff who are well-versed in the nuances of identifying, prioritizing, and mitigating obsolescence issues (see Section 5). The application of proven processes, procedures, and practices also ensures a systematic and structured approach to addressing obsolescence issues. Additionally, the success of obsolescence management can be greatly enhanced by effective cooperation among various organizations involved. For increasing effectiveness, this cooperation may be supported by a robust information system that facilitates seamless communication and coordination across the different stakeholders.

Development of a proactive programme typically takes a couple of years to reach a state where a proactive mode of the programme can be implemented using the appropriate industry tools. A proactive programme is understood to be when an NPP has a defined programme in place that ensures timely completion of maintenance activities is not impacted by obsolete items. Proactive obsolescence management can reduce the number of urgent problems, preventing production losses, outages and safety concerns. Experience has shown that where it has been implemented, stakeholders have given positive feedback on the integration into the operations and maintenance system. Further details on the advantages and disadvantages of reactive and proactive approaches are given in Refs [7, 8].

## 2.1. CHALLENGES AND SUCCESS FACTORS

Many NPPs have had challenges throughout the process of developing a recognized, formal obsolescence management programme, and ensuring accurate, reliable, up to date obsolescence data. Some of the challenges that plants faced during the development and implementation of an obsolescence management programme are listed below along with success factors for mitigating or addressing each challenge:

— Management support to implement a proactive obsolescence management programme:
  - Showing examples of close calls (i.e. instances where obsolete equipment almost led to lost time or money);
  - Showing potential benefits of a programme;
  - Highlighting industry requirements (e.g. LTO).
— Identifying the proper expertise to develop and implement the programme, and transfer the relevant knowledge:
  - Involvement in industry organizations;
  - Training for plant resources;
  - Outsourcing support.
— Transitioning from defining and establishing the programme to running the programme:
  - Continuous support from senior management for the programme;
  - Establishment of good knowledge transfer between the project team and the programme or obsolescence management team.
— Obtaining programme support from plant staff (e.g. maintenance, engineering) and the wider organization (e.g. management, support staff). The challenge may be the result of a lack of understanding of what obsolescence management means and the role each department plays in the programme. The lack of programme support can also be due to a general lack of awareness of the process. Sometimes it is difficult for plants to develop and fit the obsolescence management programme into the existing organizational structure and to get support from groups outside of the obsolescence team, especially management. Getting people to consider obsolescence in their own daily roles can be a challenge.
  - Clearly defining an interface between the obsolescence group and other organizations;
  - Obsolescence training for the whole organization for awareness of the process;
  - Management support.
— Supplier and original equipment manufacturer (OEM) willingness to cooperate and support a long term proactive programme by providing proactive information on part availability:
  - Visibility of part demand from the plants;
  - Strong supply chain management and relationships with the suppliers at the NPP.
— Awareness of all possible obsolescence solutions:
  - Understanding of industry guidance;
  - Understanding of the parts requirements;
  - Involvement in industry organizations.
— Metrics for evaluating the effectiveness of the programme and determining what good performance looks like:
  - Industry benchmarking;

- Standard performance indicators (see Refs [6, 9]).
— Having the processes in place to support the obsolescence management programme, such as an item equivalency process, a commercial grade dedication process, etc. All these processes support resolving obsolescence challenges.
— IT support and data quality within the plant asset management system is an ongoing issue that affects a plant's ability to accurately perform demand planning, obsolescence identification and prioritization. There might be difficulty obtaining support from the IT organization to transmit data to the industry IT tools.
  - Early communication on the needs and support required;
  - Performing data cleanup projects;
  - Defining data entry rules;
  - Determining what data are important for the obsolescence management programme.

## 3. LESSONS LEARNED FROM PROACTIVE OBSOLESCENCE MANAGEMENT PROGRAMMES

### 3.1. SCOPE OF THE PROGRAMME

According to TOP401, the scope of an obsolescence management programme "includes SSCs important to safety and spare parts required to maintain those SSCs but may be applied to all SSCs important for plant reliability and availability" [6]. In line with this industry recommendation, plants typically focus their programme on items important to safety, and their associated spares. Nonetheless, items not important to safety, especially those important to power generation, are typically within the scope of the programme. These classifications are typically in accordance with Ref. [10].

As will be further described in Section 3.8, industry tools can be used to identify obsolete components. This involves analysing data from the plant's data management system to determine what information is available, what is relevant for obsolescence management and the chosen tool, and what relevant information is missing from the existing plant data. Since the operating organizations' data management systems usually include a vast amount of information on every type of component and system in the plant, it is encouraged to apply filters to remove irrelevant components and systems from the obsolescence scope.

To ensure the filters applied and the scope defined is adequate, a review involving obsolescence related departments (e.g. plant engineering, maintenance) and the obsolescence engineer or process owner is helpful. This helps confirm that important systems and components are not excluded from the scope. Additionally, interfacing with relevant supply chain organizations can provide further assurance.

### 3.2. RESPONSIBILITIES

The attention, communication, adaptability, proactive problem solving, commitment to development and contribution to a team are all integral parts of a successful obsolescence management strategy. Participation in the process not only provides protection against technological disruption but supports a culture of resilience within the operating organization.

Through the feedback from Member States, it was found that the organizational unit involved in managing the obsolescence management programme varied from plant to plant and each NPP needs to adapt the programme to fit their organizational structure. There are some tasks and responsibilities that were identified to be common across different NPPs in effectively running a programme, such as the establishment of an obsolescence manager and a core obsolescence team. Overall, the number of people directly involved with the obsolescence management programme is difficult to precisely record as many people are involved on an irregular or infrequent basis.

Every participating Member State indicated the need for an obsolescence manager, responsible for the efficiency and effectiveness of the obsolescence management programme. This person will manage the screening of items for obsolescence and coordinate periodic assessment meetings to determine the need for a change or solution. This can be a dedicated position (e.g. in Sweden, Finland or Belgium), or a secondary responsibility (e.g. in the Islamic Republic of Iran).

It is important that the obsolescence manager does not make decisions alone but rather to have transparency of the process across the organization including plant engineering, procurement engineering, purchasing and maintenance. Several participating Member States have not created a dedicated obsolescence committee at their NPPs but rather use the already existing organization and work processes for handling obsolescence issues, for example, when there is a need for extra budget or to develop design change solutions or reverse engineering. Whether the plant is scheduling regular dedicated meetings with a set frequency or including obsolescence related topics in existing meetings such as the Plant Health Committee, it is important to have obsolescence discussions to allow for cooperation with all stakeholders. The main topic of discussion in these meetings is the plant 'top list', which is a prioritized list of obsolescence challenges. Decisions and actions taken by the obsolescence team are managed and recorded in accordance with the management system of the NPP.

A core obsolescence team includes engineers from diverse technical areas (e.g. mechanical, electrical, mechanical, instrumentation and control (I&C), civil) to provide a solid basis for decisions. If further technical assistance is needed, for example a design change, the issue can be delegated directly to one of the technical departments. Where possible, the obsolescence engineers will identify and prioritize issues, try to find solutions, and perform data cleaning tasks. Having a core obsolescence team also helps reduce the risk of knowledge loss.

## 3.3. PRIORITIZATION PROCESS

Prioritization has shown to be key in order to have a correct focus during multiple steps of the obsolescence management programme. Prioritization is first used when determining which items or components are proactively screened.

The prioritization criteria can be different depending on the plant needs and is determined by each operating organization. Cross functional knowledge within the organization is needed to determine and to set up the prioritization criteria. It is important for this prioritization criteria to be systematic and not driven by any single individual and their individual priorities. NPPs typically give the highest ranking to components that are important to safety. Spare parts with insufficient data are also accounted for in the prioritization process.

Some Member States (e.g. Slovenia, Sweden, Spain, Finland) use the automatic prioritization in the Proactive Obsolescence Management System (POMS), called the obsolescence value ranking (OVR) algorithm. POMS is a subscription-based programme that is owned and operated by Westinghouse Electric Company. This algorithm assigns a ranking to each component, and it can be set up to reflect the utilities obsolescence strategy. Other Member States (e.g. Belgium) utilize a plant developed scorecard to determine the priority. Several examples of approaches to prioritization and ranking are included in Section 5 of Ref. [8].

The prioritization is a living list, and it is important to keep track of the inventory balance and withdrawals because it may have an impact on the prioritization to develop a solution. Some Member States (e.g. Sweden) have integrated the inventory data with the obsolescence management database to understand the current inventory balance. Tracking precursors to obsolescence, such as increases in cost and lead-time for items purchased for plant inventory can also provide an indication of pending obsolescence.

## 3.4. IMPLEMENTATION OF SOLUTIONS

Technological obsolescence is analysed on a case by case basis. Each solution adopted will be the result of balancing technical aspects, company policies and the operating situation. It is a good practice to use a graded approach when developing solutions to obsolescence cases. Different possible solutions to obsolescence cases are shown in Fig. 1. As indicated in Fig. 1, cost and resources generally increase as solutions move up the pyramid.

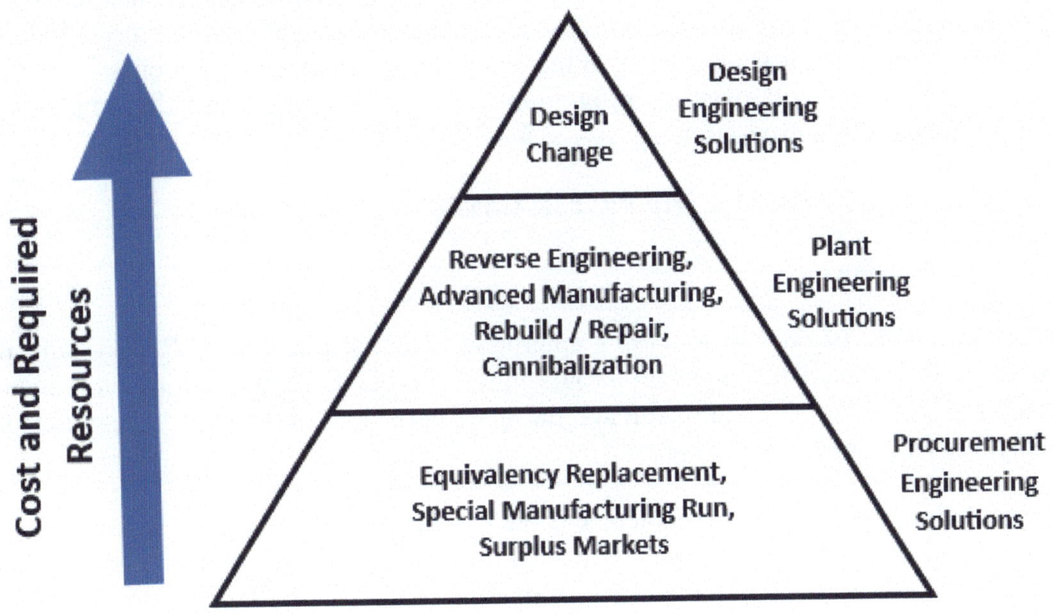


*FIG. 1. Solutions to manage technological obsolescence.*

The earlier an obsolescence case is identified the more solution paths will be available. Typically, equivalency replacement, repair, and last time buy are the most common ways of resolving obsolescence. For I&C equipment, depending on its safety functions, obsolescence issues may necessitate a design change level solution that includes modifications to the I&C architecture. There tend to be fewer obsolescence issues related to electrical equipment.

For emergent or small scale obsolescence issues, some solutions may prove to be more effective than others. For example, for simple mechanical parts the plant may choose to manufacture spare parts itself to ensure availability during an outage, given that they have this capability. Incorporating obsolescence considerations into the design change process will ensure that projects are better able to withstand changes in technology and components throughout their life cycle.

The use of advanced manufacturing techniques, including laser light measurement and additive manufacturing or 3D printing to develop obsolescence solutions is increasing. This process is new to many operating organizations of nuclear installations and may require preparation of internal processes and procedures that address additive manufacturing activities. A process for employing additive manufacturing techniques to produce spare and replacement items is presented in Ref. [11]. Section 6 provides real world examples of several obsolescence solutions identified in Fig. 1.

## 3.5. EVALUATION ALONG THE NINE ATTRIBUTES OF TOP401

IGALL TOP401 describes methods and tools for developing an obsolescence management programme. The document is organized into nine elements, similar to the organization of a standard IGALL AMP. NPPs have found that comparing their obsolescence management programme to the guidance in TOP401 can result in valuable feedback on their programme and identify areas for improvement.

Experience from Member States also shows that developing procedures and evaluating an NPP's obsolescence management programme on the basis of the nine attributes of TOP401 ensures that the resulting programme is formatted and aligned with the same systematic approach as the other programmes included in the plant's ageing management documentation. Evaluation of the nine attributes can be used to verify that the obsolescence management programme meets industry good practices and covers all significant aspects of the obsolescence management process. The evaluation can be done on an already existing programme as well as when initially implementing an obsolescence management programme.

## 3.6. DATA MANAGEMENT IN THE PROGRAMME

### 3.6.1. Data collection

The better the quality of data in an NPP's enterprise resource planning (ERP) system, the more effective the obsolescence management programme will be. Essential information includes the manufacturer and model of the equipment and the spare parts used at the plant. The goal is for NPPs to understand what is needed and when.

Regardless of the ERP system of choice, the data within these databases will be used for the identification and prioritization of obsolescence. If third party industry tools are being used for the obsolescence management programme, the information would also be transferred to the supplier or third party to identify which components or spare parts are obsolete.

### 3.6.2. Data cleanup

To address poor data quality (i.e. incomplete information on components) efforts were taken at NPPs to identify the missing information. For example, component health report analyses are conducted by some NPPs to assess the completeness of equipment and the Bill Of Materials (BOM) data for each component. Other plants use a data cleanup metric to document the progress in resolving missing information.

According to experience from the industry, actions taken to identify the missing information align with the steps defined in TOP401. Most NPPs start tracking their obsolescence challenges in parallel with implementing a data cleanup effort performed in-house or with the support of external suppliers.

## 3.7. INFORMATION EXCHANGE

### 3.7.1. Within the organization

In addition to the information that may be accessible through the plant's ERP system, obsolescence information is typically shared within the organization through reports such as system health reports, component health reports, and key performance indicator (KPI) reports.

Obsolescence challenges are typically recorded and managed through the corrective action programme. Actions are discussed and assigned to the applicable departments.

### 3.7.2. With the larger industry

Sharing information outside of the organization raises an NPP's awareness of obsolescence solutions available throughout the industry. Many obsolescence issues impact more than one NPP or utility so value can be gained by discussing solutions with other NPPs or utilities. Effective ways for plants to exchange obsolescence information are through the NUOG meetings, INUOG meetings, or direct peer-to-peer communication.

## 3.8. USE OF INDUSTRY IT TOOLS

Industry developed IT tools provide value to NPPs by providing a cost sharing opportunity for obsolescence identification and easy access to industry solutions.

Reference [7] includes a detailed list of industry tools available. The following list also provides several available industry IT tools along with a brief description of the tool. Links to many of these tools, and others, can be found on the industry tools page of the NUOG website [12].

— CMIS: engineering change and item equivalency development tool;
— EPRI Wiki: wiki site that contains technical information and resources related to nuclear procurement and supply chain;
— IRIS: operating experience of significant events;
— Nuclear Community Working Group: discussion forum for benchmarking amongst NUOG and INUOG;
— NUOG Project Center: common location for sharing NUOG documentation;
— Park Company: surplus parts database, repair solutions, obsolescence and backlog reduction tool;
— Peaks: sourcing transactions and Paragon commercial grade dedication, reverse engineering, and repair solutions;
— Peaks+: sourcing transactions, obsolescence, parts quality initiative, and inventory health;
— POP: users can report and search pending obsolescence issues, EPRI notifies potentially impacted EPRI member utilities;
— POMS: obsolescence identification and industry solutions;
— POMSV: portal for vendors to identify obsolescence issues to support the industry;
— Rapid/OIRD: locate spare parts;
— UsOne: suppliers input existing solutions, EPRI member utilities identify needed solutions; UsOne: provides quick notification of matches and facilitates development of future solutions where multiple members have a need;
— Westinghouse Parts Marketplace: inventory and surplus industry part solutions.

## 3.9. MAINTENANCE OF THE PROGRAMME

Obsolescence management is not only a separate, identified, dedicated process, but is linked to other parts of the NPP's management system and is a valuable part of total quality management. Effectiveness is followed and maintained by using KPIs and, in some NPPs, obsolescence is also part of the system health score. IGALL TOP401 provides examples of KPIs that can be used for monitoring an obsolescence management programme [6].

Effectiveness can also be measured through:

— Internal or external audits;
— Corrective Action Programme;
— SALTO missions, World Association of Nuclear Operators (WANO) and Institute of Nuclear Power Operations peer reviews, etc.;
— Benchmarking with other operating organizations individually or through Owners' Groups.

## 4. INVOLVEMENT OF THE REGULATORY BODY IN OBSOLESCENCE MANAGEMENT

There is an expectation of most regulatory bodies that NPPs will share information related to their obsolescence management; however, the specific level of details and why and when information is shared varies. Generally, the regulatory body is aware of the NPP's obsolescence management programme and supports the effort. The regulatory body may require a managed process for obsolescence, with provisions in the licensee's management system. The regulatory body may also require a demonstration of an obsolescence management programme as part of periodic safety reviews, and an obsolescence management programme is often a prerequisite for an LTO programme. Regulatory oversight is generally focused on items important to safety and components that could impact items important to safety; however, it may also include critical spares, and single point vulnerabilities (SPVs). A regulatory review of an obsolescence management programme is mostly conducted as part of a generic ageing management assessment; however, this varies from one Member State to another. Due to the unique nature of the regulations across different Member States, the following section is organized by Member State and based on feedback from participating Member States.

## 4.1. BELGIUM

The regulatory body holds periodic inspection meetings with all departments, sections and teams at NPPs, and obsolescence is one of the topics on the agenda for all entities involved in the process (Programme Engineering, Design Authority, Procurement, Maintenance, Safety, Independent Nuclear Safety Oversight and Component Health Engineers). An annual review of obsolescence arrangements is included under the framework of ageing management assessment within the Belgium Federal Agency for Nuclear Control.

Obsolescence is addressed in the safety analysis report as part of the AMP and in the periodic safety review as part of Safety Factor 3. Additionally, the programme is relevant to ageing management and LTO and is assessed against the nine generic attributes for effective ageing management in table 2 of IAEA SSG-48 [1].

The regulatory body receives the operating organization's procedures that describe the way obsolescence is managed. Item equivalency evaluation (IEE) is a recently implemented simplified process to solve obsolescence issues. IEEs can be used for simple situations with no or limited interactions between departments or sections, or no formal modification package. However, the regulatory body requires an engineering change assessment, which it reviews and approves before installation. In other cases, obsolescence issues follow the normal modification process in which the regulatory body is automatically involved. Obsolescence issues that lead to a non-conformity that is managed by a request of relaxation or exemption of a regulation is submitted to the regulatory body for approval.

Internal assessment reports from the Independent Nuclear Safety Oversight group on specific processes such as commercial grade dedication are also sent to the regulatory body, at its request.

## 4.2. FINLAND

The regulatory body requires operating organizations of NPPs to demonstrate that they have an obsolescence management programme. Obsolescence management programmes are covered briefly under ageing management in a national regulatory guide [13]. There is no strict guideline on how to demonstrate that a programme exists.

During the first trimester of each year, the operating organizations submit a follow-up report on the ageing management of SSCs to the regulatory body for information. The regulatory guide notes that "the information presented in the follow-up report may be grouped in a way deemed suitable by the licensee, for example, by equipment group and/or system" [13].

One plant has chosen to demonstrate their obsolescence management by creating an AMP that covers obsolescence management. The ageing management programme presents procedures for managing the obsolescence of SSCs. The AMP is reviewed by the regulatory body as a part of an ageing management review, which generally occurs annually. The plant has also chosen to send annually a summary of the system analyses to the regulatory body. The analyses contain summaries of spare part availability and a risk analysis of obsolescence to the system.

## 4.3. SPAIN

The Spanish nuclear regulatory body has defined internal procedures for addressing inspection activities for the ageing management of components and structures in NPPs and there is a specific chapter related to obsolescence management. This procedure establishes the requirement of having an obsolescence management programme. There is no specific regulation in this regard, but obsolescence is an issue that the regulatory body does routinely inspect and requests information about obsolescence. The regulatory body inspects the activities related to obsolescence management, preventive maintenance, and equipment reliability, as well as spare parts management for LTO.

The regulatory body issued Ref. [14] related to periodic safety reviews in NPPs, and which contains security factors that are defined based on the state of SSCs important to safety or internal operating experience. Obsolescence management needs to be addressed within these security factors.

## 4.4. SWEDEN

An operational obsolescence management programme is a prerequisite for continued operation of NPPs. The programme for obsolescence management (within managing ageing related degradation) needs to be designed in accordance with the basic concept of ageing management as described in Section 6 of SSG-48 [1] (or equivalent) and following the relevant Swedish regulations. A review of the programme is undertaken by the regulatory body on a regular basis.

## 4.5.  SWITZERLAND

The regulatory body reviews the strategies of NPPs for replacement of systems. There are clear guidelines, and associated reviews and inspections, related to replacement solutions and new installations; however, it is not referred to as obsolescence management. Obsolescence management is addressed within technological ageing monitoring in a regulatory guide [15]. This regulatory guide expects the licensee considers the availability of original manufacturer spare parts of the required quality, as well as technical support from the manufacturer. The licensees also need to consider components and systems for which neither spare part deliveries nor technical support by the manufacturer is guaranteed (referred to as discontinued components and systems in Ref. [15]). Planning the replacement of discontinued components and systems and the safety related assessment of the operation up to the time of replacement is expected, as is the handling of the possible failure of discontinued components and systems.

## 4.6.  UNITED KINGDOM

The regulatory body has requirements on the licence holder to demonstrate adequate arrangements to ensure safe operation of NPPs [16]. Of the 36 Licence Conditions (LC), the most closely related to obsolescence management include: Periodic review (LC15); Designs and Specifications (LC16); Control of design changes (LC22); and Examination, inspection, maintenance and testing (LC28). None of these LCs specifically refer to obsolescence or the required arrangement for an acceptable obsolescence management programme. However, regulatory oversight of obsolescence is included as part of general regulation of organizational arrangements (e.g. Engineering, Maintenance, Supply Chain). The regulatory body engages with NPPs to confirm that the risks associated with equipment are being properly managed, and undertake activities including:

— Site inspections and targeted interventions on plant equipment;
— Assessment of proposed plant modifications;
— Periodic safety assessments.

Design changes involving or triggered by component obsolescence would be shared with the regulatory body if associated with a significant safety system. Annual assessments of safety also include a high level summary of the current status of obsolescence management as part of ongoing assurance that plant arrangements are adequate.

The regulatory body periodically reviews the topic of obsolescence at both the plant and the fleet level. Periodic technical updates are provided to the regulatory body on themed areas. For significant equipment replacements controlled by the engineering design process the regulatory body may require review and comment on the design package before installation approval. There have been cases of regulatory interventions where the deployment of obsolescence arrangements across all plants is compared.

## 4.7.  UNITED STATES OF AMERICA

No specific regulations address obsolescence explicitly; however, regulations require that design changes are subject to design control measures commensurate with those applied to the original design. The United States of America regulatory body provides requirements and guidance on dedication of commercial grade items and services for use in NPPs. Reference [17] endorses, with identified exceptions and clarifications, Ref. [18] as a way of meeting relevant regulatory requirements. Commercial grade item dedication is often used as a solution to obsolescence issues resulting from suppliers discontinuing nuclear quality management and assurance programmes.

# 5. TRAINING AND RESOURCES FOR OBSOLESCENCE MANAGEMENT PROGRAMMES

Obsolescence management is an important issue to be considered for the safe operation of NPPs. For obsolescence management to succeed, a basic awareness of obsolescence is necessary for all involved plant personnel. To achieve this awareness, it is necessary to define the training requirements.

The competencies of an organization to manage technological obsolescence requires a highly diverse skill set of both technical and organizational knowledge. The foundation of technological obsolescence arrangements is the awareness of the steps required to manage obsolescence from identification to documented completion. All personnel involved with obsolescence need to be aware of their role and how it interfaces with other stakeholders. For the most complex obsolescence issues, technical expertise may be required from multiple groups (e.g. Engineering, Maintenance, Procurement), OEMs, or specialists. It is necessary to ensure stakeholders are suitably qualified and experienced.

The content of the training programme might differ from one State to another, or in different NPPs in the same country. However, it is valuable to define a minimum content. The following are generic items that may be included in an obsolescence training programme. NPPs can customize their programme to meet their specific needs:

- Background and introduction:
  - An explanation of the relevance of obsolescence to safe operation of the plant and the personnel involved in obsolescence management.
- Summary of the obsolescence management programme (if it exists):
  - Identification process;
  - Prioritization process;
  - Different solution paths:
    - Explanation of the most common solutions (e.g. equivalency replacement, reverse engineering, last time buy);
  - Maintenance of the obsolescence management programme.
- Roles and responsibilities:
  - Definition of the obsolescence plant committee (if it exists);
  - Explanation of plant personnel expectations.
- IT tools:
  - Internal tools (ERP system, etc.);
  - External tools (national operators' group (if it exists).
- National and international references in obsolescence management.

Additionally, general purpose training for all persons working at the NPP is suggested, as obsolescence issues might appear in any situation.

For the people directly involved in obsolescence management, participating in specific obsolescence events is an efficient way of being updated on obsolescence issues and wider operating experience. For instance, NUOG and INUOG regularly organize calls and conferences on the topic of technological obsolescence and solutions. Members are invited to present on successful solution development or their approach to obsolescence management including innovations or continuous improvement steps. In addition, the INUOG website contains several resource pages, including a list of key industry reference documents for managing obsolescence[1].

There are significant differences in the training offered, however, most Member States have some form of obsolescence training. In some NPPs, obsolescence training is provided for new employees and specific training may be provided as necessary. The training generally includes the use of the relevant IT tools. A few NPPs noted that obsolescence is part of the annual training programme, and one plant noted that it is included in the engineering qualification programme. Several noted that training occurs by participating in meetings and seminars organized by external organizations such as INUOG. The obsolescence management group may also hold informational meetings for the maintenance group and for specific individuals involved in projects impacted by obsolescence. One plant noted that periodic programme updates on solutions are delivered as newsletters and exhibitions. In addition, roadshows (directly involving external solution providers) have been used to inform wider populations. General and task-based training is delivered using the approved company process with many different media available, including approved reading of documents, operating experience briefs, coaching, computer based training, and formal classroom training.


## 6. OPERATING EXPERIENCE WITH TECHNOLOGICAL OBSOLESCENCE

As discussed in Section 3.4, unique obsolescence solutions are necessary for each situation; however, there are several overarching types of solution. These generally fall into three broad solution categories: procurement engineering, plant engineering, and design changes. Cost and necessary resources usually increase from procurement solutions to plant engineering solutions, and finally to design solutions. The following case studies provide real-world examples of obsolescence solutions and have been grouped in accordance with these solution categories.

Several of the examples below include information about the original supplier or manufacturer. This information has been included to help readers understand the specific obsolescence challenges, and in no way represents an endorsement by the IAEA of a specific product or manufacturer.

### 6.1. PROCUREMENT ENGINEERING SOLUTIONS

#### 6.1.1. Belgium: special manufacturing run/last time buy of resistance temperature detectors

ENGIE Electrabel received a notification from its supplier that the production of resistance temperature detector sensors important to safety was to be discontinued. After receiving the notification, component engineering and maintenance worked together to identify all impacted spare parts and functional locations. It was determined that the obsolescence would impact two units, more than a thousand function locations, and more than 100 spare parts.

To address the obsolescence, the INUOG community was contacted to share the supplier notification and to identify any other impacted utilities. Although no other utilities appeared to be impacted, a lot of information was provided regarding equivalent brands and models used in important to safety applications. In addition, ENGIE worked with one of their experienced buyers to understand why the

---

[1] https://www.inuog.org/docs-reference.aspx

OEM was no longer manufacturing the part. They found that the OEM was relocating their production facility from Germany to Bulgaria; however, the quality assurance department would not be relocated.

Based on this information, ENGIE developed a business case to try to convince the OEM to continue producing the component. They also prepared for a last buy. To determine the quantity, the company considered corrective replacements, component shelf life, potential life extension of units, dismantling of certain units (cannibalization), and qualified lifetime replacements. It took time to compile the list and determine the future needs. This exercise was piloted by the assistant component engineer and component expert with input from the maintenance teams, programme engineering and component supply.

After ENGIE clearly explained the situation to the OEM and developed an estimate of the necessary last buy quantity, the OEM decided to continue manufacturing the sensors for nuclear applications. ENGIE decided to still buy an increased amount of inventory, split in two phases. The first phase will contain those components that are urgently required, while the second phase will contain those components and spare parts that are less urgent.

This example shows the value of early, open and transparent communication when working with OEMs. If an adequate business case is provided, it may be possible to convince the OEM to continue supporting a component. If that is not possible, it still may be possible to agree on a last time buy opportunity that addresses the obsolescence concern. ENGIE will continue to openly engage with the OEM moving forward to ensure the sensors remain available.

### 6.1.2. Belgium: surplus procurement of radiation monitoring systems

This obsolescence success story involves the successful collaboration of the Tihange NPP and its design office, a global energy and services group, and a third-party supplier to solve a critical obsolescence issue. Multiple solution strategies were utilized to solve the obsolescence of the area radiation monitoring systems of the NPP, including surplus procurement, component repair, and commercial grade dedication.

Radiation monitoring systems installed at the facility were equipped with the Victoreën 847 detector series, which for the past 20 years had not been supported by the OEM. The detector electronics were the focal point of this project. In addition, the plant did not have the spare components required to maintain this equipment for plant operation. To address this obsolescence issue, the plant worked with a third-party supplier who researched, identified and procured like-for-like surplus components, including as-found condition analysis, testing and reporting for the surplus components. Repair plans were developed, along with commercial grade dedication plans. The third-party supplier also conducted onsite functionality testing to demonstrate that the replacement fulfils the functions of the original component.

This project was successful due to the utilization of various strategies currently recognized by the nuclear industry as proven solutions to address obsolescence, including the use of surplus materials, repair and commercial grade dedication. Combined these elements prevented costly replacement of the area radiation monitors and provided a bridging strategy to continue safe operations through future possible life extensions.

### 6.1.3. Finland: equivalent replacement of class 3 DN80 butterfly valve

Recognizing the regulatory challenges in the nuclear industry and the nuclear supply chain, Finnish licensees initiated the KELPO programme. KELPO is an ongoing procurement framework that enables the use of industrial standard components in safety class 3 and certain low risk safety class 2 applications, streamlining the procurement process for SSCs. By incorporating standardized components, the Finnish industry aims to enhance efficiency, reduce costs, and ensure compliance with Safety Standards, thereby contributing to the sustainability and safety of nuclear power generation.

While KELPO shares similarities with commercial grade dedication, it is not directly equivalent. Both approaches facilitate the use of industrial standard components in NPPs, but KELPO provides a structured and standardized approval framework, whereas commercial grade dedication is generally applied via case by case evaluations. Key differences are:

— Regulatory framework: The Finnish regulatory body has approved the KELPO process, allowing licensees to approve components systematically. Commercial grade dedication, in contrast, typically requires separate regulatory acceptance for each procurement.
— Joint approvals: KELPO enables shared approvals for product families and suppliers among Finnish licensees, reducing redundancy. Commercial grade dedication is usually limited to single components.
— Efficiency: Standardized requirements under KELPO make procurement faster and more cost-effective compared to Commercial grade dedication, which requires customized qualification for each component.

TVO has successfully applied the KELPO procedure in multiple procurement cases. One example is the replacement of a safety class 3, DN80 butterfly valve, using the item equivalency solution strategy, within the framework of the KELPO procedure. The spare part replacement work began on 18 October 2023, including the preparatory steps such as creating the valve data sheet, undergoing the internal approval cycle for the data sheet, and initiating supplier inquiries. Due to the structured nature of KELPO, and the standardization of safety class 3 components in the KELPO procedure, the order confirmation for the valves was received in just two days, on 20 October 2023. Notably, the supplier adhered to the promised delivery schedule, with the valves and all accompanying documentation arriving on schedule on 15 November 2023.

This successful case highlights the efficiency and reliability of the KELPO procedure, demonstrating how an equivalent item approach can be successfully implemented when clear requirements are defined, and suppliers are actively engaged. TVO has conducted multiple procurements following the KELPO framework, and this case serves as an example of its successful application. By integrating standardized components into nuclear procurement, KELPO enhances supply chain resilience while maintaining strict safety and regulatory compliance.

### 6.1.4. Hungary: equivalent replacement of obsolete OEM gate valve

The originally installed Russian OEM gate valves (Type 1012, DN150/133/150 PN160, butt welded design, various safety classification applications from class 2 to class 4), shown in Fig. 2, could no longer be maintained due to a number of previous repairs performed during the installed service life. In addition, there was a lack of spare parts due to the discontinuation of support by the OEM; hence, the valves were obsolete. Replacement of these obsolete valves became a necessary requirement for the plant. The obsolescence group researched and identified a new valve (Fig. 3) that met all the original design and operating parameters and end connections. It has been in reliable operation since its installation and a backup is provided.

This example demonstrates the successful use of an equivalency evaluation to provide a procurement solution. After the replacement was selected and installed, there were no changes in operation, function, or physical effects to the systems. The replacement required no additional actions other than the onsite removal and installation of the new valve. Following the replacement, an operational test (open-close) was performed on the valves in the 0–100% operating range to demonstrate functionality. The replacement valve did not change the operation of the system, and the new components had no adverse effect regarding the operation of any other items important to safety across the operating range.

*FIG. 2. Obsolete OEM Gate Valve (courtesy of B. Kocsis, Paks NPP).*

*FIG. 3. New OEM Replacement Gate Valve (courtesy of B. Kocsis, Paks NPP).*

### 6.1.5.  Sweden: special manufacturing run for butterfly valves

ESCO butterfly valves used at the Ringhals NPP were rebranded, and the OEM was not known. The supplier classified the valves as obsolete and complete valves or spares were no longer available. The plant's obsolescence team, along with the procurement department, and helped by the Chamber of Commerce, researched the item and was able to identify the OEM in Japan. The plant reached out to the OEM and was able to convince the OEM to restart production of the necessary spare parts to support a last time buy. The existing plant documentation and drawings were used to verify that the parts from the special manufacturing run were identical to the existing parts. Installation and testing were also performed at the plant to demonstrate equivalency.

This example also demonstrates the importance of reaching out to the OEM and maintaining open and transparent dialogue. Once the OEM understood the plant's needs, they were able to provide a special manufacturing run to support a last time buy.

### 6.1.6.  Sweden: throttle valve equivalent replacement

Class 4 throttle valves (model STA-D from the OEM Tour & Andersson AB) at Oskarshamn NPP were determined to be obsolete. After confirmation by the responsible maintenance engineer that the throttle valve was still needed, an assignment was created directly from the obsolescence management department to the technical department. The technical department surveyed the market to find a valve with the same specifications.

Obsolescence management received a report from the technical department. The throttle valve was replaced with a new model that is identical to the obsolete valve; therefore, the replacement could take place without plant modification.

The recommendation for the future is to replace the existing threaded valves with a flanged design. This is to facilitate maintenance, among other things, and to increase availability (threaded valves tend to leak through the threads) in the future.

The example shows how important it is for maintenance and the technical/engineering departments to communicate. Early identification of the obsolescence issue gave the technical department ample time to identify an acceptable replacement part.

### 6.1.7.   United Kingdom: replacement valve assembly

In 2022 an NPP supply chain contact was unable to procure a replacement valve assembly. An obsolescence support request was made to central contacts. Once the details of the valve and duty were confirmed various options were pursued:

— Determine if stock was available within other company locations;
— Reconfirm supplier status and seek if alternatives were offered;
— Contact likely stock organizations for any held spares;
— Contact other NPPs to see if any surplus stock was available;
— Seek alternative suppliers based on industry leads and expert contacts.

A successful outcome was identified by identifying an alternative supplier. Once the replacement possibility was identified, the component change was assessed via the normal supply chain process, enabling a recommendation for installation of the replacement assembly.

This example demonstrates the importance of maintaining strong communication and relationships with plant suppliers. Using existing contacts, the engineering department was able to quickly identify an alternative supplier.

### 6.1.8.   United States of America: surplus and equivalency evaluation for a chemical sampling system

A chemical sampling skid was obsolete due to an obsolete power supply, in which the whole skid and power supply were obsolete. The obsolescence issue was ongoing for months and resulted in manual operator chemical sampling three times a day, which was a burden to the plant. The NPP worked with a third-party supplier to identify an obsolescence solution, other than designing a new system. The supplier researched similar skids used outside the nuclear industry and was able to identify an acceptable power supply. An engineering equivalency evaluation was completed on the item to demonstrate it was an acceptable replacement. The replacement power supply was removed and tested by the supplier and shown to be adequate.

This project was successful due to the utilization of various strategies for addressing obsolescence, including the use of surplus materials from outside the nuclear industry, and equivalency evaluations. Combined these elements prevented costly redesign and replacement of the chemical sampling skids.

This example highlights the fact that possible obsolescence solutions can be identified outside of the nuclear industry.

## 6.2. PLANT ENGINEERING SOLUTIONS

### 6.2.1. Canada: replacement of electrohydraulic turbine controls

Canada deuterium–uranium (CANDU) NPPs have Parsons electrohydraulic turbine control systems that were originally installed in the early 1980s as part of the original plant design. These systems are no longer supported by the OEM and are hence obsolete.

Each system has six fixed set point modules installed, all of which are critical for precise turbine control operation of the main turbine. CANDU operating organizations partnered with a third-party supplier to address obsolescence solutions for these systems and components to ensure they remained operational. The approach utilized several solutions strategies, including repair, refurbishment, and reverse engineering in accordance with Ref. [20].

The project entailed reverse engineering or repairing multiple system components, including: the following components: servo valve amplifier, fixed speed setpoint module, fast valving module, manual valve closure module, valve closure slave module, signal interface 'C' modules, signal processing modules and feedback relay simulator.

The reverse engineering efforts were performed to meet the requirements of Appendix B of Ref. [21] and the guidance in Ref. [20]. This provided replication of the original designed components and will enable the systems to run to the end of plant life without costly digital system upgrades.

### 6.2.2. Finland: adapting inspection test plans to avoid design changes

TVO has encountered situations where changes in ownership of OEMs presented challenges in obtaining necessary documentation for spare parts. In some cases, the new manufacturer was unable or unwilling to provide all documentation required by the TVO inspection test plan, rendering the spare part effectively obsolete. Consequently, design change projects were initiated to replace these spare parts. However, upon reviewing inspection test plans, opportunities were found to avoid time consuming design changes. For instance, there was a case in which carefully examining the test requirements for quick exhaust valves identified specific tests that the current manufacturer couldn't perform (e.g. pressure and leakage tests). Subsequently, the inspection test plans were adjusted to conduct these tests at TVO's facilities while ensuring compliance with regulatory requirements.

This proactive approach not only spared the plant from time-consuming design changes but also emphasized the importance of adaptability and thorough scrutiny in addressing challenges within the spare part process.

### 6.2.3. France: replacement of programmable logic controllers

There are almost 600 programmable logic controllers (PLCs) installed at ORANO La Hague. Most of these PLCs were installed in the 1990s and are now obsolete; however, there are too many to migrate all of them to a new model. The plant intends to operate for another 30 years and has developed a two-tiered plan to address the obsolete PLCs. Tier 1 involves identifying and migrating the most important PLCs, while Tier 2 involves repair and reverse engineering.

A review was undertaken to identify the PLCs that drive the most important functions of the plant, and a shortlist has been established. The plant intends to replace those PLCs with up-to-date PLC models over the next several years. The old PLCs dismantled during these updates, will be recovered, reconditioned, and put into stock to increase the number of spare parts. The replacement of an old PLC is expensive and difficult, and there is a risk that the new PLC will also become obsolete. The plant plans to cooperate closely with the provider of the new PLC design to extend its availability.

Despite these migrations, many original PLCs will remain on the site. For these PLCs, the plant intends to ensure operability through repair, stock improvement, and reverse engineering. For the repair, ORANO has hired a subcontractor to repair electronic equipment and moved them within the site to increase productivity and effectiveness. As equipment will stay in service far longer than in other industries, there is the opportunity to collect decommissioned assets. In some cases, the decommissioned cards are stored separately and only repaired when the regular stock is reduced. This approach requires managing two stocks.

ORANO is also studying the principal cards of the original PLCs with the aim of being able to reverse engineer the PLCs. The goal is to have prototypes and be ready to produce the main cards if needed. One challenge of this approach is negotiating the use of intellectual property that belongs to the OEM.

It is challenging to maintain the necessary programming skills for obsolete PLCs, and this challenge will likely increase as knowledgeable programmers continue to leave the industry.

The existing PLCs are obsolete, and actions needed to be taken to keep them operating for 30 more years. ORANO has determined that using multiple obsolescence solution strategies is the best approach. This ongoing project will use a multi-solution approach including equivalent replacement, repair, and reverse engineering.

### 6.2.4. Spain: replacement of obsolete boric acid pump

Asco NPP requested a third party to assist in finding a solution to obsolete ASME N-stamp boric acid pumps for a pressurized water reactor (PWR). These pumps were manufactured by an OEM that no longer supported their ASME programme, including any support for this pump model. Asco did not want to modify the piping and preferred a 'drop-in' solution with a like-for-like design. To accomplish this, the supplier decided to use a reverse engineering solution strategy along with an equivalency report and commercial grade dedication.

The NPPs procurement engineering organization worked closely with the third-party supplier to ensure equivalent pump components could be developed and to ensure proper comprehensive and robust equivalency documentation was developed.

The supplier was given access to a spare boric acid pump. The supplier technicians travelled to Asco, disassembled the pump (see Fig. 4), and began the reverse engineering process. All parts were laid out piece by piece. The supplier utilized a state of the art Laser Coordinate Measuring Machine to capture the geometry of the castings, such as, the impeller, casing, power frame, and stuffing box cover. For the machined parts (i.e. shaft, sleeve, and hardware), the supplier used calibrated micrometres. The technicians deployed were well versed in pump parts and focused on the key aspects of the design. Once the geometry was captured, the supplier developed a 3D model of the replacement piece, as shown in Fig. 5.

The material was specified in the BOM. Some materials were no longer available on the market. For those materials the supplier chose replacement materials with a Technical Design Justification to ensure the material would perform to the design and its intended function. The supplier took design control over the entire pump and provided the following:

— ASME Design Report.
— A new BOM and drawing with the supplier part numbers for every part.
— Seismic Report.
— Equivalency Report:
  • Geometric evidence that the geometry to be used for manufacture matches the OEM design;
  • Actual drawing with sizes of the sketch for machined parts at the plant;
  • 3D model of OEM piece vs. the new supplier piece overlapped with geometric deviation analysis.
— Failure Modes and Effects Analysis.
— Hydraulic Pump Test verifying identical performance (see Fig. 6).
— Commercial Grade Dedication Package.

The supplier successfully provided an identical new pump replacement with full quality assurance documentation to support equivalency. A total of nine pumps were manufactured and all are performing well. This addressed the NPP's obsolescence issue.

*FIG. 4. Obsolete pump disassembled at plant. Ready for reverse engineering (courtesy of F. Salman, Hydro Inc.).*

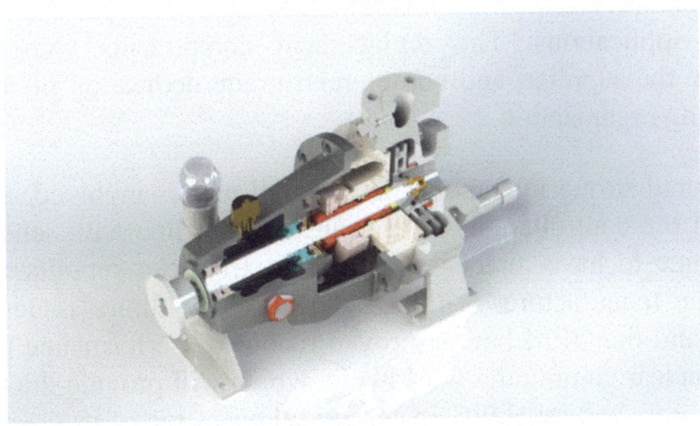

*FIG. 5. 3D model cross section of the reverse engineered and manufactured pump (courtesy of F. Salman, Hydro Inc.).*

*FIG. 6. Hydraulic pump test being performed to verify identical performance (courtesy of F. Salman, Hydro Inc.).*

### 6.2.5.   United Kingdom: reverse engineering of actuator controller

It was determined that the HVAC system included aged and obsolete actuator controllers with limited spare parts. Using the remaining sample of the original spare, a design was reverse engineered, established, prototyped, and tested. The design was tested for key functionality and validated as a suitable replacement. The design has been fully approved for manufacture and use.

## 6.3.   DESIGN SOLUTIONS

### 6.3.1.   Canada: replacement and redesign of scan line controllers

This ongoing obsolescence replacement project is part of a long term life extension project for a multi-unit NPP in Canada and consists of replacing 645 scan line controllers with a customized series controller. There was intense collaboration between the utility, its contractors, the supplier, and the manufacturer as this project involved customizing the supplied controller to plant specifications. To ensure a successful result, the utility worked with the supplier and the manufacturer to modify the basic specifications of the controller to create a 'plug-and-play' replacement design for the NPP. Additionally, the supplier worked with the manufacturer to implement additional modifications to support unique plant applications. The replacement components were environmentally and seismically qualified by the supplier, and commercial grade dedication plans were developed and submitted to the operating organization.

Due to the cooperative and strong interaction between all parties involved, a feasible solution was developed and many of the controllers have been installed in line with scheduling. The operating organization will continue to have direct representation from the supplier whose interaction and communication with the manufacturer will help address any future issues that may arise. The replacement controller and the OEM have a proven record of performance not only in the current application but in the nuclear community worldwide, which will provide the operating organization with confidence of durability and reliability in successful operation of its reactors.

### 6.3.2.   United Kingdom: replacement of motor protection relays

In 2016, several motor protection relays for items important to safety were confirmed as obsolete. The existing system was based on P&B Gold type electromechanical relays. Age-related degradation was causing a decrease in the reliability of the relays, leading to spurious activation and reduced availability of associated equipment.

It was determined that a modern microprocessor-based relay system could be used for the replacement, which involved a technology change. The updated technology would provide ongoing spares availability, manufacturer support and reliable replacements until the planned end of asset life. The replacements were justified within the engineering change process. The final project deliverables included qualification, safety case demonstration, functional specifications, a testing strategy, factory acceptance tests, seismic qualification, statistical testing, and installation support, including engineering change completion.

# REFERENCES

[1] INTERNATIONAL ATOMIC ENERGY AGENCY, Ageing Management and Development of a Programme for Long Term Operation of Nuclear Power Plants, IAEA Safety Standards Series No. SSG-48, IAEA, Vienna (2018).

[1] INTERNATIONAL ATOMIC ENERGY AGENCY, Safety of Nuclear Power Plants: Commissioning and Operation, IAEA Safety Standards Series No. SSR-2/2 (Rev. 1), IAEA, Vienna (2016).

[2] INTERNATIONAL ATOMIC ENERGY AGENCY, Management of Ageing and Obsolescence of Instrumentation and Control Systems and Equipment in Nuclear Power Plants and Related Facilities Through Modernization, IAEA Nuclear Energy Series No. NR-T-3.34, IAEA, Vienna (2022).

[4] INTERNATIONAL ATOMIC ENERGY AGENCY, Procurement Engineering and Supply Chain Guidelines in Support of Operation and Maintenance of Nuclear Facilities, IAEA Nuclear Energy Series No. NP-T-3.21, Vienna (2016).

[5] INTERNATIONAL ATOMIC ENERGY AGENCY, Ageing Management for Nuclear Power Plants: International Generic Ageing Lessons Learned (IGALL), Safety Reports Series No. 82 (Rev. 2), IAEA, Vienna (2024). https://doi.org/qhb8

[6] INTERNATIONAL ATOMIC ENERGY AGENCY, Technological Obsolescence Programme Version 2023, IGALL TOP401, IAEA, Vienna (2023). https://gnssn.iaea.org/NSNI/PoS/IGALL/SitePages/Home.aspx

[7] INSTITUTE OF NUCLEAR POWER OPERATIONS, Obsolescence Guideline, Nuclear Utility Obsolescence Group International Nuclear Obsolescence Group, INPO NX-1037 (Rev. 3), Atlanta (2018).

[8] ELECTRIC POWER RESEARCH INSTITUTE, Plant Support Engineering: Proactive Obsolescence Management, Program Implementation and Lessons Learned, EPRI 1019161, Palo Alto (2009).

[9] INTERNATIONAL NUCLEAR UTILITIES OBSOLESCENCE GROUP, Guideline for Establishing Obsolescence Indicators, INUOG-2018/002 (2018).

[10] INSTITUTE OF NUCLEAR POWER OPERATIONS, Equipment Reliability Process Description, INPO AP-913 (Rev. 8), Atlanta (2018).

[11] ELECTRIC POWER RESEARCH INSTITUTE, Additive Manufacturing to Support Acquisition of Spare and Replacement Items, EPRI 3002023735, Palo Alto (2022).

[12] NUCLEAR UTILITY OBSOLESCENCE GROUP, NUOG Industry Tools (2024), https://www.nuog.org/industry-tools.aspx

[13] FINISH RADIATION AND NUCLEAR SAFETY AUTHORITY, Ageing Management of a Nuclear Facility, Guide YVL A.8, STUK (2019).

[14] SPANISH NUCLEAR SAFETY COUNCIL, Periodic Safety Reviews of Nuclear Power Plants, Security Guide 1.10 (Rev. 2), CSN, Madrid (2017).

[15] SWISS FEDERAL NUCLEAR SAFETY INSPECTORATE, Periodic Safety Review for Nuclear Power Plants, Guideline for Swiss Nuclear Installations ENSI-A03, ENSI (2014).

[16]     UK OFFICE FOR NUCLEAR REGULATION, Licence Condition Handbook, ONR (2017).

[17]     US NUCLEAR REGULATORY COMMISSION, Dedication of Commercial-Grade Items for use in Nuclear Power Plants, Regulatory Guide 1.164 (Rev. 1), NRC, Washington DC (2024), https://adamswebsearch2.nrc.gov/webSearch2/main.jsp?AccessionNumber=ML24038A310

[18]     ELECTRIC POWER RESEARCH INSTITUTE, Plant Engineering: Guideline for the Acceptance of Commercial-Grade Items in Nuclear Safety-Related Applications, Revision 1 to EPRI NP-5652 and TR-102260, EPRI 3002002982, Palo Alto (2014).

[19]     INTERNATIONAL NUCLEAR UTILITY OBSOLESCENCE GROUP, Key Industry Reference Documents (2024), https://www.inuog.org/docs-reference.aspx

[20]     ELECTRIC POWER RESEARCH INSTITUTE, Guidance for the Use of Reverse-Engineering Techniques, Revision 1 to EPRI TR-107372, EPRI 3002011678, Palo Alto (2018).

[21]     US NUCLEAR REGULATORY COMMISSION, 10 CFR Part 50, Appendix B, Quality Assurance Criteria for Nuclear Power Plants and Fuel Reprocessing Plants, Washington DC, Latest Edition.

## EXPERIENCES OF MEMBER STATES ON DEVELOPMENT AND IMPLEMENTATION OF PROACTIVE OBSOLESCENCE PROGRAMMES

During IGALL Phase 5 (2020–2021), participating Member States were requested to provide information on their obsolescence management programmes. Member States were asked the questions in Section A–1. Nine Member States provided answers during the IGALL survey or during the development of this publication. The responses provided by the Member States are included in Sections A–2 to A–13. As noted, most of these responses were provided in 2020/2021. In some areas, the responses have been updated to reflect that status at the beginning of 2025; however, this is not the case for all responses.

## A–1. QUESTIONS TO THE MEMBER STATES

### A–1.1. Description of the NPP/utility

### A–1.2. Experience in development of an obsolescence management programme

### A–1.3. Description of the proactive obsolescence management programme

*A–1.3.1.   Scope of the programme*

— Data collection (e.g. from plant asset management system);
— Availability of replacement equipment and spare parts from manufacturers;
— Efforts to identify missing information, data cleanup (document review, walk-downs, inputs from suppliers).

*A–1.3.2.   Responsibilities*

— For planning and implementation;
— Which organizational units like engineering, maintenance, operations and work planning units, plant senior management are involved and how their involvement is managed;
— Use of an obsolescence committee and experts;
— Who is coordinating with supply chain organizations and how the coordination is accomplished.

*A–1.3.3.   Prioritization process*

— What criteria are used (safety, demand, quantity in stock, failure history; reliability of structures, systems and components (SSCs); work order information; stock history; spare parts with insufficient data);
— What priority categories are defined;
— Who does the prioritization and how the prioritization is accomplished.

*A–1.3.4.   Implementation of solutions*

— Design change;
— Reverse engineering, rebuild or repair, spares from other components;
— Evaluation of equivalency, special manufacturing run, procurement of surplus items from market.

*A–1.3.5.   Evaluation along the nine attributes*

IGALL TOP401 describes what methods/tools can be applied under the specific attributes. In this section the experience on using these methods or tools is described.

*A–1.3.6.   Data management in the programme*

How information and data exchange take place within the organization and with the industrial organizations (supplier, manufacturer, etc.).

*A–1.3.7.   Use of IT tools in implementation*

— Databases and software used;
— Experience with the use of IT tools.

*A–1.3.8.   Evaluation and maintenance of the programme*

— Description of experience with the use of the programme;
— How operating experience is fed back into the programme;
— How effectiveness is maintained (e.g. use of self-assessment, external reviews, performance indicators).

## A–1.4. Involvement of the regulatory body

Licensees' experience with regulatory reviews, approvals, and inspections.

## A–1.5. Training

How understanding of obsolescence management of those involved is developed and maintained.

## A–1.6. Operating experience with technological obsolescence

Description of real technological obsolescence operating experience cases.

## A–2. RESPONSE FROM BELGIUM (ENGIE ELECTRABEL)

## A–2.1. Description of the plant(s)

ENGIE Electrabel owns and operates the two NPP sites in Belgium (Doel and Tihange) and provided information from both sites.

## A–2.2.   Experience in development of an obsolescence management programme

In 2017, the appointment of the first fleet Equipment Qualification & Obsolescence manager within ENGIE Electrabel has led to an increased awareness and focus on obsolescence management. Another key element in the development of an obsolescence programme was the appointment of site obsolescence managers for the development and coordination of the programme on both sites.

Before 2017, the management of obsolescence was done sporadically and rather reactively. Obsolescence was at that time managed by the component engineers, which were part of the

Maintenance Department. After the appointment of the first site obsolescence manager a process mapping exercise was performed to analyse the bottlenecks, pros and cons, and lead times of the existing methodology for treating obsolescence. Potential improvement actions were determined and grouped in a performance improvement plan. Active engagement in industry work groups such as INUOG gave valuable insight on the experience from other operating organizations who had already implemented a programme. A strategy for proactive obsolescence management was developed based on the IAEA TOP401 document. The aspects of 'detect', 'prioritize', 'solve' and 'monitor' form the basis of the programme.

In 2018, a customized IT tool was developed to manage the proactive screening of items/components, help prioritize obsolescence issues and to have a transparent overview of ongoing obsolescence files. The IT tool also made it possible to visualize the impact of obsolescence at a plant and system level. System engineers are able to analyse the impact on functional locations and assess the overall thread of obsolescence from a system point of view. Component engineers are able to analyse which component types are impacted by obsolescence and which obsolescence files are ongoing in order to mitigate the impact.

Developing a proactive programme, including the development and roll-out of a customized IT tool, took about two years for the Doel nuclear site. The Tihange nuclear site started the development of an obsolescence programme one year later but both sites had synchronized processes by the end of 2020.

## A–2.3. Description of the proactive obsolescence management programme

### A–2.3.1. Scope of the programme

The SAP ERP software is used within ENGIE Electrabel as the master system for plant asset, work, inventory, and document management. During the analysis of the former obsolescence programme, it was identified that SAP was not transparent and flexible enough to be used for obsolescence management. The customized IT tool that was developed extracts daily all relevant data from SAP and visualizes it in a very transparent way. The SAP data are used to make calculations and estimations of (for example) the stock-out dates of the items and components in the system. Since equipment data and the Bill of Materials (BOM) are not always complete, the IT tool leverages as much data available from supply chain requests, purchase orders, maintenance work orders, PM orders, etc. to reconcile missing data.

The scope of the obsolescence programme considers all components, including items (piece parts) that have been created as a spare part in the SAP system. Since equipment without any spare parts in their BOM are excluded from the scope of the proactive obsolescence programme, efforts are made to clean up these data. During a component health report analysis, a component expert assesses the completeness of equipment and BOM data for each component type. When critical spare parts are missing in the BOM a corrective action is determined to create new spare parts and/or complete the BOM data.

### A–2.3.2. Responsibilities

The site obsolescence managers are responsible for the efficiency and effectiveness of the programme. Proactive screening of obsolescence is piloted by the obsolescence manager, as well as the coordination of periodic 'assessment meetings' where the need for a solution is determined.

Proactive screening is performed with the cooperation of the Purchasing Department, who contact suppliers to screen the availability of their items/components. Tractebel Engineering, ENGIE's external design engineering bureau, also performs obsolescence screening during 'type test audits' of suppliers of items/components important to safety. These audits are performed every three years.

During assessment meetings the need and urgency for a solution is (re)evaluated. During these meetings the component engineers and experts (maintenance department) are consulted. On site there are four senior component engineers (1 electrical, 1 instrumentation and 2 mechanical engineers) who each have one junior assistant engineer. The component engineers are in charge of a team of senior component experts who typically have years of experience in plant maintenance. From time to time, maintenance engineers and maintenance technicians are also involved during these assessment meetings. Since component engineers and experts are responsible for ageing management, technical analysis of failures, technical expertise, etc., obsolescence is just a small part of their daily job.

For each domain an obsolescence committee is hosted by the obsolescence manager. The component engineers however are responsible for validation of the calculated priority score for each obsolescence file. They also manage the ranking of all obsolescence files and commit to milestones and deadlines. Programme engineers are also active participants to the obsolescence committee. They are responsible for the equipment qualification phase of an obsolescence file. Engineers from Tractebel Engineering participate as design engineers and are each responsible for the technical management of several obsolescence files. Purchasing and Warehousing, Supply Chain, and Supply Chain Quality Assurance also actively participate in the obsolescence committee at the Tihange nuclear site. At the Doel nuclear site they do not systematically participate, only upon specific invitation.

Operations and Work Planning Departments are not systematically involved in the obsolescence programme. For specific obsolescence files where reverse engineering is being performed, they are consulted whenever in situ conditions of the items or components need to be verified during outage or normal operation.

On a fleet level there are two Equipment Qualification and Obsolescence Managers. Once a month an equipment qualification and obsolescence technical committee is hosted where the issues in the domain of equipment qualification and obsolescence are discussed. Both processes are discussed together as they have shown to be strongly interrelated. The fleet supplier and goods qualification manager is also an active participant and is considered as a valuable source of input for the domain of supplier audits and qualification. During the technical committee, the site senior managers of Component Engineering (Maintenance), Supply Chain and Programme Engineering are present, as well as the obsolescence managers from both sites. The senior Tractebel Engineering manager is also an active participant. From time to time a manager from Laborelec, ENGIE's research and material test facility, is invited.

### A–2.3.3. *Prioritization process*

Prioritization has shown to be key to ensuring the correct focus during multiple steps of the obsolescence programme. Prioritization is first used when determining which items/components need to be proactively screened. Also, a scorecard is used to determine the priority of all obsolescence files.

To determine which items/components need to be proactively screened for obsolescence, the IT tool automatically identifies those items/components that are important to safety and are projected to run out of stock before the end of life of the NPPs. Those items/components that have been delivered on site within the last 12 months are taken out of scope as they are considered to have a lower chance of unforeseen obsolescence since there was a recent contact with the OEM. This scope can be manually enlarged by the obsolescence manager to take into account those items/components that are not important to safety but critical for plant availability.

All obsolescence files are automatically scored in the customized IT tool using an objective scorecard. The scorecard considers the level of safety classification, qualification level, calculated stock-out date, current stock and existing non-conformities due to the lack of spare parts. This automatic score card can be manually adjusted by the component engineer. When stock levels change, the score cards are automatically recalculated.

*A–2.3.4. Implementation of solutions*

The first goal of the proactive obsolescence screening is to increase the number of last buys to reduce to impact of obsolescence and bridge towards the end of life of the NPPs. When a last buy is not possible the future needs for both sites is assessed, and a special manufacturing run is requested by the purchasing department. Even though efforts are made to increase the number of special manufacturing runs, quantities are typically too low to convince the manufacturers. When stakes are high, INUOG members are contacted to verify whether other impacted operating organizations are interested to join forces.

If possible, a long term rebuild/repair strategy is applied. Therefore, the necessary long term repair contracts are being negotiated. The most applied solution for obsolescence is replacing the obsolete component by an equivalent design. A design change is typically considered as a last resort. When, after the scoping of an obsolescence file, a design change is unavoidable, the file is transferred toward the Engineering Projects Department. Reverse engineering is performed sporadically and is typically managed as a separate project.

A modification file is currently used as the primary tool for design control due to the lack of an IEE process. However, a special 'type modification' can be used, by which an equivalent component is proactively investigated, so it can be replaced when corrective maintenance is necessary.

In 2019, an equipment qualification and obsolescence project was launched by fleet engineering to search for and challenge alternative processes for obsolescence and equipment qualification. Development of IEE process was within the scope of this project. As well as the optimization and standardization of the process for reverse engineering. The goal of this project was to establish clear procedures, guidelines and templates by the end of 2020. Each of these deliverables were tested in cooperation with both sites during real life test cases. Guides and procedures are implemented on both sites and applied with the support of the fleet team.

*A–2.3.5. Evaluation along the nine attributes*

1. Scope of the technological obsolescence programme:

The primary focus of the obsolescence programme is SSCs important to safety. However, items and components important to plant reliability and availability are also addressed.

2. Proactive technological obsolescence actions to minimize and control obsolescence:

The obsolescence programme is piloted by two fleet equipment qualification and obsolescence managers and two obsolescence managers responsible for a specific nuclear site. There are sufficient resources (organizational) available to proactively address obsolescence. The resources that participate in the management of obsolescence have expertise on the SSCs being treated. Unsolved obsolescence issues are reviewed periodically by a dedicated obsolescence committee on the sites. The top common obsolescence issues between the sites are discussed during a periodic fleet committee.

The proactive approach is implemented on multiple levels of the programme, during identification, prioritization, and solution development. The method for the prioritization of items/components that will be proactively screened for obsolescence is based on historical consumption data. If item/component/equipment data appear incomplete, corrective actions are determined to complete these data. After items/components are identified as obsolete, a scorecard is used to prioritize the obsolescence issue. All ongoing obsolescence files are ranked in accordance with their score. Obsolescence files are managed trough predefined milestones and an estimated due date.

3. Detection of technological obsolescence indicators:

No online monitoring of 'early indicators' is performed within the obsolescence programme of ENGIE Electrabel. The obsolescence managers receive a bi-monthly newsletter from the fleet supplier manager, containing general supplier news, supplier organizational changes, critical suppliers, credit risk information, etc. This newsletter can trigger the obsolescence manager to launch a proactive obsolescence screening. This is, however, not done systematically at the moment.

4. Monitoring and trending of technological obsolescence:

Annual self-assessments are performed at a fleet and site level to assess the effectiveness of the programme. During the self-assessment, the evolution of the metrics over the past 12 months is discussed. When necessary, corrective actions are determined. At the Doel site, metrics are available online in the IT tool for obsolescence management. The most important metrics are communicated monthly to the senior site management. After the implementation of the IT tool at the Tihange site (Q3 of 2020), online metrics are available for both sites. Currently, comparable metrics for both sites can be monitored from a fleet point of view.

5. Mitigating technological obsolescence:

Reactive and proactive obsolescence issues are managed by obsolescence files. Each file has predetermined milestones and a due date. If, at a plant level, an adverse event occurs due to obsolescence and lacking spare parts, this is managed in accordance with the ENGIE policy for reporting of non-conformities. The author of a non-conformity report needs to indicate the cause of the non-conformity. 'Obsolescence' is one of the available cause codes, which makes it possible to analyse trends in the number of non-conformities due to obsolescence. Corrective actions with due dates are determined for each non-conformity report. In the IT tool for obsolescence management, non-conformity reports and their corrective actions can be linked to the obsolescence file. In this way the committed due date of the corrective action is visible and is used to correctly determine the due date of the obsolescence file.

6. Acceptance criteria:

There are clear and documented acceptance criteria for the obsolescence programme. The adherence to the acceptance criteria is monitored by a metric.

7. Corrective actions:

If at a plant level an adverse event occurs due to obsolescence and lacking spare parts, this is managed in accordance with the ENGIE policy for reporting of non-conformities. The author of a non-conformity report needs to indicate the cause of the non-conformity. 'Obsolescence' is one of the available cause codes, which makes it possible to perform trending of the number of non-conformities due to obsolescence. Corrective actions with due dates are determined for each non-conformity report.

8. Operating experience feedback:

Internal operating experience is shared between both sites during the equipment qualification and obsolescence technical committee. ENGIE Electrabel also has actively participated in the INUOG community since 2017 to share experience and learn from other operating organizations. In addition, ENGIE also promotes participation in other related workgroups and industry forums (EPRI Joint Utilities Task Group, Curtiss-Wright/NUGEC annual EQ technical meeting, etc.). Continuous improvement in the domain of obsolescence management is strongly promoted. In 2019–2020 ENGIE Electrabel performed a benchmark visit at ANAV in the form of a WANO Member Support Mission. ENGIE Electrabel also participated in an EPRI 'Obsolescence Assessment' requested by Kernkraftwerk Leibstadt AG. Both missions gave valuable insight in the obsolescence processes of peer operating organizations.

9. Quality management:

The programme is implemented in accordance with national regulatory requirements.

*A–2.3.6.  Data management in the programme*

Proactive obsolescence screenings, supplier feedback and obsolescence file data are stored in the customized IT tool. It was chosen to store information about obsolescence in a centralized database with a direct link to the impacted spare parts. Most critical data (for example, spare part obsolescence codes) are also stored in SAP. Since some of the information is confidential, it is possible to restrict access to certain documents to a limited group of users.

*A–2.3.7.  Use of IT tools in implementation*

A customized IT tool was developed to facilitate the management of obsolescence. Other tools available in the industry were also investigated. Since the environmental and seismic qualification of equivalent equipment is also managed during an obsolescence file, ENGIE Electrabel needed a strong tool for file follow-up. There wasn't a module available in the industry that could fulfil the needs of file follow-up. The customized IT tool makes it possible to have a transparent overview of the progress of an obsolescence file by using milestone and due date monitoring. Tasks can be linked to each phase of an obsolescence file. Each actor can also generate its own task list, where tasks are ranked in accordance with the obsolescence file score and task due date.

The advantage of the IT tool is that all information about obsolescence is centralized, from detection of obsolescence until the closure of an obsolescence file. Procurement also uses the same IT tool to manage and monitor supply chain files. Requests for the creation of a new spare part are launched in the same database. This makes it possible to have full traceability from the detection of the obsolescence, until the delivery of the new equivalent item on site. Another advantage is that all information about reactive detection of obsolescence is also available in this IT tool.

*A–2.3.8.  Evaluation and maintenance of the programme*

The experience shows that the IT tool contributes to centralization of obsolescence data and leads to transparency in the process for all actors involved. Operating experience with the IT tool is gathered by the site obsolescence manager. Small modifications to the software are managed by a yearly software maintenance contract. After one year of implementation of the IT tool, a software self-assessment was performed with all the users.

Detection of obsolescence

The experience with proactive screening of obsolescence is that it takes a lot of effort to contact and challenge OEMs. Support from an experienced buyer with technical background is important. When discovering that a 'last buy' or 'special manufacturing run' is possible, the workload of correctly determining the order quantity cannot be underestimated. Since the proactive screening has led to several last buys there is also a non-negligible impact on the inventory value. It is important to proactively inform the inventory manager about the obsolescence strategy because typically the objective of inventory management is to reduce the inventory value towards the end of life of the NPP. Performing last buys will have an impact on inventory value but will lead, in the long run, to a cost saving by avoiding equipment qualification and/or design modifications.

Prioritization of obsolescence

The experience with prioritization has shown that a score card is an effective tool to prioritize obsolescence issues. ENGIE Electrabel uses a scorecard with five parameters. Even though some utilities use a score card with many more parameters it was decided to limit the number of parameters to only those with complete data. In this way all scorecards are automatically calculated without any errors.

## A–2.4. Involvement of the regulatory body

The obsolescence programmes of both sites were presented in 2019 to the regulatory body during the Annual Meeting on the Ageing Management Programme of ENGIE. The final report remarked that the obsolescence management of both sites was significantly improved and that positive results could be already observed on the key performance indicators related to the ageing management. The regulatory body recognized the efforts made by ENGIE to improve the Ageing and Obsolescence Management Programme, to make it compliant with international good practices, to develop adequate tools for its management, and to assign the necessary resources to the process. The Federal Agency for Nuclear Control informed ENGIE that they will continue to pay special attention to this topic in the coming years.

## A–2.5. Training

There are local site procedures available that describe obsolescence management in general. A fleet governance procedure developed in the framework of the equipment qualification and obsolescence optimization project is available.

Also, specific guidelines are available to work with the IT tool. Most of the knowledge is transferred by on the job training and demonstration sessions. On the job training is available for system and programme engineers on how to work with the IT tool and interpret the available data, obsolescence files, and to comprehend the chosen mitigation strategies.

In addition, specific sessions on different approaches to resolving obsolescence cases (reverse engineering, assessment of item equivalence, dedication of commercial products) are organized to train all staff potentially concerned.

## A–2.6. Operating experience with technological obsolescence

In 2018, ENGIE Electrabel received a notification from a supplier that they were going to discontinue the production of resistance temperature detector (RTD) sensors important to safety because of a relocation of the production facility. This obsolescence impacted both nuclear sites, more than a thousand functional locations and more than 100 spare parts. After several communications between the supplier and one of ENGIE's experienced buyers, ENGIE was offered the opportunity to perform a last buy. An action plan was developed that consisted of five main steps:

1. Determine the impact

The current facility where the RTDs where designed and manufactured had a valid certificate of authorization where ENGIE could buy all items as a basic component. Losing one of those manufacturers always has a big impact on the supply chain. Both sites were impacted by this obsolescence. ENGIE identified all impacted spare parts and functional locations together with Component Engineering and Maintenance.

The INUOG community was contacted to share the supplier notification and to see which other operating organizations were impacted. It seemed that ENGIE Electrabel was the only operating organization in INUOG that were using these specific RTDs. There were two other operating organizations that had used similar RTDs before, but they had already replaced them by an equivalent model. Through INUOG, ENGIE received a lot of information about equivalent brands and models that were used by other operating organizations in items important to safety.

2. Contact the OEM

ENGIE teamed up with an experienced buyer to understand the obsolescence notification. The reason for leaving the nuclear industry seemed to be due to the relocation of their production facility from Germany to Bulgaria. The current quality assurance department would not be relocated. However, on ENGIE's request they agreed to temporarily move the quality assurance department to Bulgaria to support a last buy.

3. Develop a business case

Together with the support of their buyer ENGIE tried to develop an interesting business case to convince the supplier to maintain the production of important to safety RTDs in Bulgaria. One of the ideas was to group all purchase orders over a period of time to make it more efficient for the supplier to deal with the purchase orders.

4. Prepare last buy

To determine the order quantities, both sites determined their future needs. This exercise took into account: corrective replacements, component shelf life, potential life extension of units, dismantling of certain units (cannibalization) and qualified lifetime replacements. It took some time to compile the list and determine the future needs. This exercise was piloted by the assistant component engineer and component expert with input from the maintenance teams, programme engineering and component supply.

5. Develop a back-up plan

In case the last buy could not be performed, a back-up plan was identified. Through INUOG members ENGIE received information about several other vendors of RTDs important to safety.

At the time of this publication, the supplier was willing to provide continuous support and was no longer intending to stop the manufacture of sensors for nuclear applications. Nevertheless, ENGIE still decided to buy an increased amount of inventory. The manufacturing was split up in two phases. The first phase contains those RTDs that are urgently required by ENGIE. The second phase contains those RTDs and spare parts that are less urgent. ENGIE learned that transparent communication with the supplier gave the supplier insight in ENGIE's future component needs. It also allowed the supplier to support a business case to keep manufacturing RTDs important to safety. Further discussions will be necessary to agree upon the way of cooperating in the future.

## A–3. RESPONSE FROM CANADA (BRUCE POWER)

### A–3.1. Description of the plant(s)

Bruce Power operates the Bruce Nuclear Generating Station in Ontario, Canada.

### A–3.2. Experience in development of an obsolescence management programme

Bruce Power established an obsolescence management programme in 2012 after internally identifying an area for improvement for obsolescence. Bruce Power received WANO Strengths for Obsolescence Management at Bruce A and Bruce B in 2014 and 2016, respectively.

In 2016, Bruce Power received an agreement for long term operation (LTO) of the NPP until 2064. To support continuous improvement, the obsolescence procedure has been revised multiple times to take into account improvements in prioritization, alignment with organizational changes in the supply chain and the plant, and performance monitoring oversight.

Challenges experienced through development of the obsolescence programme include:

— Reorganizations in engineering and supply chain resulting in changes in roles and responsibilities within the obsolescence programme governance;
— Prioritization algorithm updates based on changing plant priorities;
— Resource availability in supply chain and engineering to support proactive obsolescence resolutions.

### A–3.3. Description of the proactive obsolescence management programme

*A–3.3.1. Scope of the programme*

Bruce Power collects data from the plant asset management system and POMS to support proactive obsolescence identification. The following processes are leveraged:

— Previous Procurement History: If the part was purchased within the past five years, it is noted as not obsolete in the obsolescence management programme.
— Vendor Substitution Forms: Through the procurement process, these allow for the OEM to identify replacement parts.

— POMS industry replacement solutions: This includes OEM, alternate manufacturer, and operating organization solutions.
— POMS spare parts availability status: As part of POMS confirming obsolescence of a part, they also ask the OEM if they still support spare parts.

Bruce Power has developed a data cleanup metric for SPV equipment. A focused project effort has been executed to improve the manufacturer/model data quality by approximately 25% since 2017. In general, design document reviews and walkdowns are performed to resolve missing information.

### A–3.3.2. *Responsibilities*

Bruce Power's obsolescence management programme is performed within the Engineering Organization. Within the programme, there are obsolescence process coordinators that coordinate, engage, and interact with the rest of the organization. The coordinators hold reoccurring working solution meetings with Equipment Performance Engineering to proactively address top obsolescence issues. The obsolescence management programme assists with identifying obsolescence issues on the work plan, which is then resolved through the permit request process.

The equipment reliability peer group is an internal multi-organizational team of senior managers, which has direct oversight of the obsolescence management programme process and performance. Obsolescence key performance indicators are presented semi-annually to the Station Plan Health Committee and the Nuclear Safety Review Board.

### A–3.3.3. *Prioritization process*

An expert panel reviews the action plans' prioritization approximately every three months.

### A–3.3.4. *Implementation of solutions*

Bruce Power takes a graded approach to the solution path for resolving obsolescence based on cost and complexity of the solution. Due to the recent life extension, replacing an obsolete part with a commercially available part is preferred and prioritized over a repair strategy. For the replacement strategy, the replacement part is evaluated through the item equivalency screening process to determine if a short or long form item equivalency can be performed or if a design change is needed.

### A–3.3.5. *Evaluation along the nine attributes*

Bruce Power benchmarked off the TOP401 document in 2014/2015 and incorporated the relevant content into the obsolescence management governance. Notable content incorporated into the governance was:

— Metrics and monitoring;
— Advancing the portfolio to serve as a 'health' document;
— Supply chain resilience trend analysis;
— Identification of obsolescence events;
— Process health;
— Ageing of equipment;
— Flagging storage of obsolete parts and work completed on obsolete parts;
— Early warning indicators of obsolete parts.

### A–3.3.6. *Data management in the programme*

No response provided.

### A–3.3.7. *Use of IT tools in implementation*

Bruce Power has been using the Obsolescence Manager module in POMS to identify obsolescence risks and document and track obsolescence solutions through all stages of resolution. In 2020, Bruce Power developed internal software to manage the overall programme including identification, prioritization and solution development. All data are stored on a server that allows integration with other applications and organizations. Microsoft Power BI reports and dashboards have also been developed to monitor and communicate programme performance.

### A–3.3.8. *Evaluation and maintenance of the programme*

Bruce Power has developed KPIs based on industry best standards to monitor and trend performance. KPIs include 1) Overall availability status of components (Obsolete, Not Obsolete, Unknown, Unusable) by criticality, 2) Completed Obsolescence Solutions, 3) Obsolescence Solutions Ready for Execution, 4) Equipment is Unusable Manufacturer/Model Information.

## A–3.4. Involvement of the regulatory body

No response provided.

## A–3.5. Training

Bruce Power has developed computer based training on the obsolescence programme, which is linked to engineering qualifications.

## A–3.6. Operating experience with technological obsolescence

No response provided.


## A–4. RESPONSE FROM FINLAND (FORTUM)

## A–4.1. Description of the plant(s)

Fortum operates the Loviisa NPP in Loviisa, Finland. Loviisa is a two-unit pressurized water reactor (PWR) site.

## A–4.2. Experience in development of an obsolescence management programme

Loviisa NPP developed and implemented an obsolescence management programme between 2016 and 2017, together with POMS implementation. The biggest challenge was getting people involved and to consider obsolescence matters in their own roles. Another big challenge was improving data quality and effective data cleanup.

## A–4.3. Description of the proactive obsolescence management programme

### A–4.3.1. *Scope of the programme*

Data are collected from the plant asset management system where they are transferred to the POMS system. Loviisa put some efforts into identify missing information, including:

— Hire summer trainees to collect missing data from field;
— System engineers walkdowns;
— Improve processes how to maintain data quality in plant asset management system.

### A–4.3.2. Responsibilities

System engineers and spare part engineers follow and identify issues, the Plant Health Committee prioritizes the issues, design engineers create solutions (e.g. technical evaluations, qualifications, design changes), and the obsolescence manager, along with the obsolescence coordinator, follow the obsolescence management process and reports quarterly to the Plant Health Committee and plant executive team.

### A–4.3.3. Prioritization process

The prioritization criteria used are: safety, criticality classification, SPV, demand, quantity in stock, failure history; reliability of SSCs; work order information; stock history; and spare parts with insufficient data. Priority categories are based on the OVR in POMS and the Plant Health Committee's prioritization values.

### A–4.3.4. Implementation of solutions

After implementing the obsolescence management programme, the solution range has been slightly wider than earlier.

### A–4.3.5. Evaluation along the nine attributes

IGALL TOP401 was one of the main sources used when the plant developed the obsolescence management programme.

### A–4.3.6. Data management in the programme

No response provided

### A–4.3.7. Use of IT tools in implementation

The computerized maintenance management system database used at Loviisa is Maximo Nuclear. POMS is used for obsolescence management. Many self-developed excel database tools are used, for example to conduct data cleanup, track stock, etc.

### A–4.3.8. Evaluation and maintenance of the programme

Effectiveness is followed and maintained by using programme KPIs in POMS. Obsolescence is also part of system health scores. Internal audits are performed periodically, and a report is provided quarterly to the Plant Health Committee. The obsolescence guideline is evaluated and updated at least every four years.

## A–4.4. Involvement of the regulatory body

The regulatory body makes reviews periodically and the obsolescence management programme is part of those reviews.

## A–4.5. Training

Basic obsolescence training is part of the individual training programme for new employees and those who have changed their position. Specific obsolescence training is provided when needed.

## A–4.6. Operating experience with technological obsolescence

No response provided.

## A–5. RESPONSE FROM FINLAND (TVO)

### A–5.1. Description of the plant(s)

TVO operates the Olkiluoto NPP in Finland. The Olkiluoto site has two boiling water reactors (BWRs) commissioned in 1979 and 1982, and one PWR commissioned in 2023.

### A–5.2.  Experience in development of an obsolescence management programme

TVO has been an active member of INUOG, and the support received from the group has been valuable in developing TVO's obsolescence management programme. The implementation of POMS has been a key step in transitioning towards a more proactive approach to obsolescence management, enabling the identification and resolution of issues before they affect safety.

Data quality challenges related to stock item information in the plant asset management system initially limited the efficient use of POMS. To address this, TVO conducted a POMS pilot in 2024, focusing on data cleanup and refining the way manufacturer and component type information is structured in the system. Previously, stock item data were spread across multiple fields, which complicated the use of POMS. The pilot resulted in a streamlined process in which manufacturer and component type data are now provided in dedicated fields, improving data accuracy and making it more compatible with POMS requirements. The pilot also provided valuable insights into additional data requirements from manufacturers. As a result, TVO updated its stock item creation guidelines, ensuring that new data entries meet the necessary quality standards. This has significantly improved the quality and usability of data within POMS, enabling successful identification of stock items and allowing POMS to determine their obsolescence status more effectively.

At the same time, TVO is developing an equipment reliability process, which is a risk-based approach to asset management, prioritizing equipment to ensure long term reliability. Since obsolescence is a key risk factor affecting asset performance, integrating obsolescence management into the equipment reliability framework strengthens both programmes.

### A–5.3.  Description of the proactive obsolescence management programme

*A–5.3.1.  Scope of the programme*

Manufacturers and suppliers are often able to suggest replacement items for obsolete spare parts. There is currently a lot of variation in the quality of stock item data. Efforts are taken to identify and fill missing information and to correct stock item information in the plant asset management system. Stock items that are not identified in POMS, will be corrected in a prioritized order. There will be an obsolescence flag system implemented in the plant asset management system.

### A–5.3.2.   Responsibilities

System engineers, spare parts coordinators and a spare part process steering group follow obsolescence issues. Obsolescence specialists, the purchasing organization and others, identify and flag obsolescence issues. An expert panel with the aid of a prioritization IT tool, prioritizes issues. Design engineers create solutions (e.g. technical evaluations, qualifications, design changes).

Obsolescence specialists follow the obsolescence management process and report to the Plant Health Committee.

### A–5.3.3.   Prioritization process

Action plans have a prioritization IT tool, in which the action plans are assigned a priority class from 0 to 5. Classification is based on maintenance classification (criticality) and stock availability. Class 0 basically means that the plant is already late with an important spare part issue.

### A–5.3.4.   Implementation of solutions

Different solutions are evaluated in a pre-study for the action plan to solve an obsolescence issue. The most common solution is a spare part replacement with an equivalent spare part.

### A–5.3.5.   Evaluation along the nine attributes

No response provided.

### A–5.3.6.   Data management in the programme

Obsolescence status and action plans are reviewed in the system health report process. An annual component report (component responsible report) reviews obsolescence issues, this way obsolescence issues are also considered in programmes related to component locations. Most critical obsolescence action plans are followed in the spare part steering group monthly. All obsolescence action plans are followed, coordinated, and centralized in a custom IT tool. Engineering disciplines have their team meetings covering obsolescence action plans. Approved obsolescence solutions are stored in the document management system Kronodoc.

There will be a stock item obsolescence status flag implemented in the plant asset management system. The stock items in the system have a link to action plans to solve the obsolescence issue.

POMS is used for identifying obsolescence and sharing obsolescence solutions.

### A–5.3.7.   Use of IT tools in implementation

The plant uses a customized plant asset management system. POMS is used as an IT tool to identify obsolete stock items and to find potential replacement spare parts. Prioritization of obsolescence issues is done with a custom IT tool and expert panel. Stock management is done in IFS (an ERP system).

*A–5.3.8.  Evaluation and maintenance of the programme*

The plant has a Plant Health Committee where the topic of obsolescence is covered periodically.

Obsolescence is part of the system health score. System health utilizes POMS KPIs.

## A–5.4. Involvement of the regulatory body

Obsolescence management is part of a larger entity of ageing management, and as a part of it, it is covered in a facility-specific periodic inspection programme.

## A–5.5. Training

Obsolescence related processes are documented. Training for the tasks related to obsolescence management is given by the respective organization where the work is done. There are instructions for IT tools. A basic introduction to obsolescence is part of ageing management training.

The operating organization is implementing a flag system for obsolescence status for stock items in the plant asset management system. The operating organization plans to actively investigate the obsolescence status for stock items. There will be instructions for these tasks and IT tools for the work.

## A–5.6. Operating experience with technological obsolescence

No response provided.


## A–6. RESPONSE FROM THE ISLAMIC REPUBLIC OF IRAN (BUSHEHR NPP)

### A–6.1. Description of the plant(s)

Information was provided for the Bushehr NPP.

### A–6.2.  Experience in development of an obsolescence management programme

Obsolescence management has been challenging at the NPP due to several factors, including:

— Mixed technology in design and construction;
— Use of several standards and regulations for SSCs;
— Prolonged construction delay;
— Issues following the accident at the Fukushima Daichi NPP.

### A–6.3.  Description of the proactive obsolescence management programme

*A–6.3.1.  Scope of the programme*

The scope of the obsolescence management programme covers all standards, regulations, design technical documents, procedures and guidance relating to the functions of Bushehr-1. For the technological obsolescence programme, the scope is reduced in such a way that covers just part and component obsolescence related to the replacement of components and spare parts. Arrangements are made for availability of replacement and spare parts at the time of purchasing and concluding contracts.

Regarding missing information, provisions are made to recover the data and information in different sources including walkdowns, reverse engineering, contacting the supplier, searching the internet, localization and adaptation.

### A–6.3.2. *Responsibilities*

The operating organization has the prime responsibility to implement the obsolescence management programme, as well as the ageing management programme. To implement this programme, several units including maintenance, safety and engineering, and ageing management working groups are involved. Coordination with the supply chain organization is made through the predefined official means and intermediate mechanism.

### A–6.3.3. *Prioritization process*

Regarding obsolescence management activities, acceptance criteria is provided in accordance with the objectives of safety, compliancy with the current licencing basis, reliability and updating accident and safety data, standards and regulations. To fulfil obsolescence management activities, the prioritization is done based on a graded approach.

### A–6.3.4. *Implementation of solutions*

The solutions for obsolescence issues are obtained by several means including reverse engineering, redesign, rebuild, repair, replacement by spare parts and providing similar components from supply chain manufacturers.

### A–6.3.5. *Evaluation along the nine attributes*

No response provided.

### A–6.3.6. *Data management in the programme*

Information within the organization and the industry is exchanged through technical reports, inspections and walkdowns, and internet platforms.

### A–6.3.7. *Use of IT tools in implementation*

All the related components and parts are classified in a database. Traditional IT tools are used when necessary.

### A–6.3.8. *Evaluation and maintenance of the programme*

The operating experience is gathered by considering the indicators reported in annual safety reports, accident reports, daily performance measures, etc.

## A–6.4. Involvement of the regulatory body

The outputs of audits, inspection and assessment of safety provided by the regulatory body is used in this regard.

## A–6.5. Training

Through training, the experience of other national or international organizations and States can be shared and the lessons implemented.

## A–6.6. Operating experience with technological obsolescence

Technological obsolescence was experienced during the construction and operation of Bushehr-1. Due to a prolonged construction delay, and mixed technology and different standards and components, storage, supply, and maintenance (condition monitoring) of spare parts has also been a challenging issue.

## A–7. RESPONSE FROM SLOVENIA (NUKLEARNA ELEKTRARNA KRŠKO)

### A–7.1. Description of the plant(s)

Nuklearna Elektrarna Krško (NEK) operates the Krško NPP in Krško, Slovenia. Krško is a single 727 MW PWR unit, jointly and equally owned by the electric power utilities of Slovenia and Croatia.

### A–7.2.  Experience in development of an obsolescence management programme

With more than 30 years of operation, and plans to consider operation for 60 years, procurement may be a challenge for Krško. Consolidation of vendors and suppliers, or their complete departure from the nuclear industry, reduces supply sources and decreases competition. Krško is a single unit operating in accordance with American regulations, codes and standards, which means the plant has to address the challenges associated with ordering small volumes of parts in a narrow, targeted market. In case of obsolete items, additional time and costs are involved for engineering (identification of identical or alternative replacement, equivalency evaluations, design reviews and modifications) and qualification efforts.

NEK has an annual investment programme addressing improvements, replacements, licensing requirements, power upgrades, and safety upgrades. This investment programme has led to the replacement of many major components, which has helped minimize the impact of obsolescence on the NPP. The investment programme maintains a conservative philosophy when developing improvement or replacement plans, and when possible proven solutions are applied. Unless not otherwise possible, first of a kind designs are avoided, and like-for-like solutions are prioritized. The equivalency evaluation process addresses the adequacy of these solutions.

### A–7.3.  Description of the proactive obsolescence management programme

#### A–7.3.1.  Scope of the programme

There has been a large effort to document every plant component in the asset management system. The master equipment component list is a fully controlled database reflecting the actual as-built plant status. Updating the database is an integral step in the plant modifications and work order processes. The master equipment component list is a connected document database with components.

Work orders are issued on components; upon completion, all configuration changes are applied if spare components were issued from stock (serial numbers, drawings, part numbers). A preventive maintenance database has been established with scheduling and work generation template modules. The same database and modules are used for in-storage maintenance.

Complete equipment list, stock, work orders and maintenance plans are sent to POMS and updated every six months.

NEK has a continuous investment programme (most of the major components have been replaced) thereby reducing the vulnerability to obsolescence. There is a programme in place to have spare parts readily available in stock for components important to safety. Preventive maintenance reports can provide data for future maintenance challenges, giving maintenance personnel information to order stock to accommodate long lead times.

A corrective action programme has been established to identify and record (even anonymously) all discrepancies in the field. A corrective actions committee reviews and assigns actions to applicable organizations.

The vendor contact programme in POMS is a welcome addition to gain information from suppliers and vendors. A top 100 list is reviewed by an obsolescence committee (system engineers, maintenance, procurement, engineering).

Every work order is potentially one of the tools used to update database with new information, when discrepancies are observed in field. Feedback is recorded in collection plans.

*A–7.3.2. Responsibilities*

The maintenance manager is the owner of the obsolescence management programme. The preventive maintenance coordinator reports to the maintenance manager, maintains the database and communicates with other organizations. Specific responsibilities include the following:

— The Expert Commission for Maintenance Efficiency Control (SK NUV) is responsible for considering proposals for action plans to address the issue of equipment obsolescence.
— The Technical Director is responsible for considering and approving the proposals of SK NUV for solving the obsolescence of equipment and implementing solutions within the action plans.
— The Engineering Manager is responsible for coordinating and effectively managing activities related to the obsolescence of equipment (project changes, technical substitutability evaluation, dedication, procurement engineering support). For new systems/components long term support is a requirement and evaluation criteria.
— The Purchasing Manager is responsible for coordinating and effectively managing activities related to the procurement of critical goods, as well as coordinating communications and monitoring the status of suppliers and manufacturers of critical equipment.
— The Maintenance Manager is responsible for the implementation of corrective actions and the revision of preventive maintenance programmes in accordance with the recommendations of the SK NUV. He is responsible for the management, coordination and supervision of activities related to the monitoring of the POMS database.
— The head of each individual maintenance discipline is responsible for the implementation of corrective actions and the revision of relevant preventive maintenance programmes, in accordance with the recommendations of the SK NUV. This person is responsible for the timely initiation of requests for ordering critical goods that appear on the obsolescence list and coordinating the implementation of prescribed actions in the implementation of solutions (technical substitutability evaluation, dedication, modifications). This person also participates

in the work of SK NUV in the identification, prioritization of the action plan and coordination in the implementation of the approved solution.

— The Responsible Engineer (initiator) is responsible for:

  • Monitoring the status of the BOM and ordering the necessary spare parts in a timely manner;

  • Preparing a technical substitutability evaluation for obsolete equipment that is not in the procurement process, in accordance with the procedure.

— The Head of Engineering Support is responsible for coordinating engineering support in the process of purchasing obsolete equipment and actions defined for the implementation of approved equipment obsolescence solutions (dedication, technical substitutability evaluation, defining items and changing statuses).

— The Procurement Support Engineer is responsible for carrying out all technical activities necessary for the procurement of obsolete equipment in accordance with the procedure.

— The Preventive Maintenance Coordinator is responsible for the management, coordination and supervision of activities related to the monitoring of the POMS database and is an intermediate link between the application and the technologists. The Coordinator monitors the POMS database and searches for solutions for NEK obsolete equipment within the application.

— The Modification Manager is responsible for providing support in approving project changes.

— The Modification Engineer is responsible for implementing the design changes.

— The Operations Manager is responsible for providing support and coordinating the work of the System Engineers.

— System Engineers are responsible for identifying critical goods on the basis of quarterly reports on the state of the systems, setting priorities, preparing an action plan for monitoring the obsolescence of critical goods and for engineering evaluation of activities.

— The Contract Administrator is responsible for communication with suppliers based on information from Engineering Support.

Progress is monitored within the system reports prepared quarterly and discussed at the SK NUV committee.

Procurement engineering coordinates with supply chain organizations.

*A–7.3.3.  Prioritization process*

The OVR score in POMS is calculated as shown in Table A–1. The first seven lines provide component level information, while the remaining lines provide information from work orders and stock levels.

TABLE A–1. NEK PRIORITIZATION PARAMETERS IN POMS

| PARAMETER | 5 POINTS | 3 POINTS | 1 POINT | WEIGHT |
| --- | --- | --- | --- | --- |
| Criticality | CC1 | CC2 | NC | 10% |
| Environmental Qualification | N/A | Yes | N/A | 10% |
| Regulatory Guide 1.97 | N/A | Yes | N/A | 10% |
| SPV | Yes | N/A | N/A | 10% |
| Man/Model Install Count | $\geq 20$ | $\geq 10$ and $< 20$ | $< 10$ | 3% |
| Component Quality Class | SR | AQ | N/A | 10% |
| Seismic Category | N/A | I | II | 7% |
| Months to Next Work Order | $< 3$ | $\geq 3$ and $< 6$ | $\geq 6$ and $< 12$ | 10% |

| | | | | |
|---|---|---|---|---|
| Historical WO Count | $\geq 5$ | $\geq 2$ and $< 5$ | $\geq 1$ and $< 2$ | 10% |
| Task Type | CM - Corrective Maintenance | DM - Deficiency Maintenance | PM - Preventive Maintenance | 5% |
| Lead Time | N/A | $\geq 360$ | $\geq 180$ and $< 360$ | 5% |
| Stock Level | $< 1$ | $\geq 1$ and $< 4$ | $\geq 4$ and $< 6$ | 10% |

The Preventive Maintenance Coordinator exports the Top 100 list from POMS and distributes it to the heads of individual disciplines, who are in charge of maintaining the installed equipment. In cooperation with the responsible engineers, a priority list is determined, which is then entered into the Living List. The top 100 list is made based on the OVR, which is part of the POMS application. Components from Living Lists are entered into the corrective actions programme. SK NUV prioritizes and assigns actions and recommendations to responsible disciplines (engineering, modifications, procurement).

### A–7.3.4.   Implementation of solutions

All implementation solutions (i.e. procurement, plant engineering, design changes) may be considered and could be applicable. Existing evaluations from RAPID and POMS are considered first. Internal reverse engineering is avoided in principle, but third-party companies with proven solutions are considered. Additional corrections are made for different component specifications/requirements. When there is no identified solution available the modification process is initiated for a design change.

### A–7.3.5.   Evaluation along the nine attributes

No response provided.

### A–7.3.6.   Data management in the programme

Information and data exchange take place via standard communication channels such as telephone, fax, and email. Oracle Enterprise Business Suite with custom modifications and tools provides communications channels via forums, requests, revision forms and workflows.

The corrective action programme is used to record and manage obsolescence issues among all other issues on site (hardware, software, procedures, documents, drawings, etc.). Actions are discussed and assigned to applicable departments.

For the POMS application, the IT department exports required data from the database and uploads the data via a secured ftp link to a dedicated folder in the POMS server. Additional communication through feedback is possible in the POMS application.

### A–7.3.7.   Use of IT tools in implementation

The following databases and software have been used:

— Oracle Enterprise Business Suite containing modules for Enterprise Asset Management, Inventory, Purchasing, Projects, Finance;
— Corrective Actions Programme;
— Rapid (Curtiss-Wright);
— POMS (Westinghouse).

RAPID and POMS have also been a valuable addition for recognizing and mitigating obsolescence issues.

*A–7.3.8.    Evaluation and maintenance of the programme*

Regular self-assessment and external reviews are performed.

## A–7.4. Involvement of the regulatory body

Regular regulatory reviews are performed. Nearing end of life additional reviews were and will be performed for the extension period.

## A–7.5. Training

Regular training is performed on many different topics providing new information from industry and regulatory bodies.

## A–7.6. Operating experience with technological obsolescence

As mentioned, NEK has a yearly investment programme addressing improvements, replacements, licensing requirements, power and safety upgrades. This minimizes the occurrence of technological obsolescence.

Reverse engineering and 3D printing has been used in the past to address obsolescence and this was recognized as a best practice in the nuclear industry.

In a recent maintenance rule committee meeting, the obsolescence of an RM25 pump (radiation monitoring) was discussed, and replacement was initiated. This proactive approach allowed NEK to recognize the issue in advance and ensure spare parts for 2–3 years.


## A–8. RESPONSE FROM SPAIN (ANAV)

## A–8.1. Description of the plant(s)

Asociación Nuclear Ascó Vandellòs (ANAV), operates three NPPs in Spain. Two PWRs at the Asco site, and one PWR at Vandellos.

## A–8.2.    Experience in development of an obsolescence management programme

ANAV has been solving obsolescence problems for years with many different solution paths (design changes, equivalence analysis (with and without design change), surplus market, special manufacturing runs). Although ANAV was solving obsolescence issues, there was not a formal obsolescence programme in place. The Spanish regulatory body and international interest groups gave the input to ANAV to create a working group to develop an obsolescence programme. The Technical Services director created a working group of individuals from units that regularly face obsolescence problems including the technical director, procurement engineering, plant engineering, design engineering, maintenance, IT, procurement, and inventory management.

The product of that working group was a report with the guidelines to establish the obsolescence programme at ANAV, which had four main areas:

— Procedures: issue obsolescence procedures and review the obsolescence related procedures (obsolescence committee, reverse engineering, etc.);
— IT Adaption: obtain software used in the industry and develop interfaces with the ANAV system;
— Database update: cleanup data in the components and spares databases;
— Training and communication: training was provided for obsolescence related personnel and the programme was communicated to the rest of the company.

## A–8.3.  Description of the proactive obsolescence management programme

### A–8.3.1.  *Scope of the programme*

The obsolescence programme is defined for the components that are included in the equipment reliability process, especially for those that are assigned safety classes C1, C2 or C3. That also means that spares for those components are in the scope of the process.

ANAV has a warehouse where spare parts and components are stored. The stock level is enough to have necessary parts for planned maintenance and includes a margin for unplanned maintenance.

If the stock level is lower than the reorder point (or there is an order for a planned task) and the same part/component cannot be found in the market during the purchasing process, it is called reactive obsolescence (even if there are spares in the warehouse) and leads to a purchasing issue that is generally solved by equivalence analysis or a different procurement solution.

When ANAV started receiving results from POMS, it became clear that the database needed to be cleaned up. Two projects were started, which are ongoing. A components database cleanup where the plant components were updated by reviewing design documents, conducting walkdowns, review modification documentation, etc., and a spares database cleanup.

### A–8.3.2.  *Responsibilities*

Technical Services are the group responsible for the plan but the implementation runs from Logistic Services. The technical director, procurement engineering, plant engineering, design engineering, maintenance; IT; procurement; and inventory management are all involved in obsolescence management.

There is an obsolescence committee in each site, which meets five times per year and the tasks are: review top list; review site new identifications; review obsolescence problems; review past meeting notes and actions. The committee consists of plant engineering manager; procurement engineering manager; head of mechanic maintenance; head of electric maintenance; head of I&C maintenance; material purchasing manager; obsolescence coordinator; procurement engineering technical support manager, and any other individual who is required to attend.

In ANAV, there is a logistics division where supply chain (procurement) and procurement engineering are two different sections but within the same division. Since the supply chain (material purchasing manager) is represented in the obsolescence committee, it is aligned with obsolescence interests.

### A–8.3.3.  *Prioritization process*

Obsolescence issues are prioritized based on the following:

— Components importance:
  - Safety qualification;
  - ER qualification;
  - SPV Yes/No?;
  - Environmental qualification;
  - Plant footprint.
— Component demand:
  - Frozen work orders;
  - Work order priority;
  - Spares consumed in last 5 years;
  - Future demand of spares.
— Components availability:
  - Lead time of spares;
  - Time without spare for work orders.

The plant engineering manager defined the matrix, and the matrix was issued by getting information from industry (other NPPs, NUOG, EPRI, etc.).

### A–8.3.4. *Implementation of solutions*

Design change as a solution of obsolescence involves a change of system, component and sometimes a part. This type of solution is implemented when a full system is obsolete, and it will affect a large number of components or parts. Equivalence analyses are sometimes documented as design changes when the change has an impact on the configuration of the plant.

An equivalency evaluation is typically done for parts and a fit for function analysis is performed. Sometimes the replacement is direct and does not require a fit for function analysis. In this case, a design change has to be performed in order to document the change in the plant.

Reverse engineering is a process to design and manufacture a new component by using an existing component or part as a model. The new component is analysed as an alternative component. Analysing components before discarding them is a good practice. The removed component may be able to be repaired or rebuilt, then used as a replacement, or the removed component cannot be repaired but may be cannibalized for parts.

Sometimes manufacturers have removed equipment from the market due to low demand, but they will consider a special manufacture run, if the order is large enough. Considering the number of spare parts needed until the end of the estimated life of the plant or system can make orders large enough to justify a special manufacturing run.

Procurement of items from the surplus market is typically performed as a bridging solution. There may be opportunities to purchase surplus items when a similar plant is decommissioned, or when an NPP performs a design change, and their spares are no longer necessary.

### A–8.3.5. *Evaluation along the nine attributes*

ANAV performed an evaluation of the obsolescence process with the nine attributes of TOP401. The result was that the ANAV process meets most of the guidance in that document. The points that are not met already are being addressed by actions in process. Once completed, the ANAV process is expected to meet TOP401 guidance. The following summarizes key portions of the ANAV obsolescence process along the TOP401 attributes:

1. Provides a clear definition of what is obsolete. ANAV has the same guidance as TOP401 for what is considered obsolete.

2. Establishes concepts that need to be clearly defined within the organization and provides a methodology that the industry recommends. ANAV has both organizational responsibilities and a methodology aligned with TOP401 guidance.

3. Some circumstances, such as purchasing problems or information from industry forums (i.e. INUOG), may identify a potential obsolescence issue in the future. When ANAV identifies these indicators, action is taken to mitigate the problem.

4. Monitoring and assessing the obsolescence process is necessary to determine its performance. ANAV has pending actions to set KPIs. An assessment has not been performed yet.

5. Solution and mitigation actions. Solutions were explained above and in ANAV those solutions are used as required after an analysis.

6. Acceptance criteria have to be developed, and the criteria related to performance indicators. Currently ANAV does not have defined acceptance criteria. It will be defined in accordance with the indicators, once they are implemented.

7. Corrective actions need to be tracked. There are a set of obsolescence actions that are reviewed and tracked in the obsolescence committees.

8. Operating experience is essential. ANAV takes part in industry groups (e.g. NUOG, EPRI) to be aware about industry news, suppliers news, share site experiences, etc.

9. Quality management. All procedures are issued in accordance with regulatory requirements and after each committee meeting notes are signed and sent to official document storage.

*A–8.3.6. Data management in the programme*

ANAV uses the POMS service that includes updated parts information from suppliers and manufacturers in a database. An automated process makes information from this database available to the ANAV obsolescence database. In addition, when a purchasing process cannot be done because a supplier identifies a part as obsolete, there is a process for solving purchasing issues. Supply chain and procurement engineering also maintain contacts with suppliers.

*A–8.3.7. Use of IT tools in implementation*

ANAV has an obsolescence database for stock codes and plant equipment that is updated with POMS information. ANAV also uses industry recommended software tools, including POMS and RAPID.

It is necessary to clean up the existing parts data in order to ensure complete and accurate information when developing an obsolescence management programme. Once the cleanup is done, proactive management could begin. It was also necessary to make some changes in the IT system of the company in order to accommodate the obsolescence databases and to inform the fields.

*A–8.3.8. Evaluation and maintenance of the programme*

After training and communication actions, and after some obsolescence committee meetings, the process started to provide results, and already some bulk purchase orders have been implemented to get spares easily, and modifications were prioritized to ensure no problems with obsolete components.

When a significant obsolescence problem is solved, it is presented in an INUOG meeting, and solutions are regularly updated in POMS.

An official assessment has not been completed yet because the process is new, nevertheless ANAV has been tracking improvements to be made.

### A–8.4. Involvement of the regulatory body

The regulatory body has performed inspections to review how ANAV deals with obsolescence problems. In 2012, the regulatory body initially issued actions for ANAV to develop an obsolescence programme.

### A–8.5. Training

A part of the programme was training and communication. The technical services director provided some communication sessions to managers. In addition, the annual training programme included the obsolescence process, software tools and the corporate obsolescence identification tool, in order that the people involved in the obsolescence process are informed and trained.

### A–8.6. Operating experience with technological obsolescence

A supplier informed ANAV that some of the relays used in the plant were going to be obsolete. After some discussion, the supplier accepted to run a special manufacture with the obsolete models. ANAV estimated all relays to be used until end of plant life and issued a large purchase order. This order also included relays for other Spanish NPPs.


## A–9. RESPONSE FROM SWEDEN (RINGHALS)

### A–9.1. Description of the plant(s)

Vattenfall operates Ringhals NPP, which has two PWR units and is located on the west coast of Sweden. Earlier reactors (one BWR and one PWR) at the site are permanently shut down.

### A–9.2. Experience in development of an obsolescence management programme

The plant started to develop a proactive obsolescence programme in 2015, and it was an action taken because of a pre-SALTO mission, which had pointed out that the programme in place was reactive and only focused on existing obsolescence issues. The pre-SALTO mission peer to peer contact provided a lot of help and information about improving the existing programme.

Taking part in INUOG and NUOG meetings was very important when forming a programme. The member operating organizations are very open and helpful in sharing programmes, lessons learned, good practice and co-operation in the area of obsolescence. Sharing of this knowledge and peer to peer connection saved a lot of time and was beneficial for setting up a successful programme. However different organizations, reporting and management structures makes it necessary to adopt an obsolescence programme to the existing organization.

POMS is used as a tool for information about known or upcoming obsolescence and this gives the obsolescence team the opportunity to start the development of an obsolescence solution before the obsolescence impacts the plant.

To understand upcoming demands of spare parts and components, the data condition in the plant register is essential (e.g. BOM, maintenance plans, model and manufacturer, etc.). Incomplete and missing data increase the risk of missing obsolescence issues and will affect the prioritization of obsolete items. The POMS database helps by correcting information; however, the operating organization still has to make efforts to correct data and enhance the plant register to enable full understanding of the obsolescence situation. To achieve this, support is needed from outside the obsolescence group, otherwise there is a risk that the obsolescence team will focus on data cleanup instead of solving obsolescence issues.

The programme is within the maintenance organization, and it is managed with an obsolescence manager. Resources involved are obsolescence engineers in close cooperation with material planner and spare part engineers. System health and maintenance engineers are essential for correct prioritization of obsolescence issues.

TOP401 was used as a reference when forming the plant programme and it gives good support in what the obsolescence programme is supposed to achieve. It took two years to develop an obsolescence programme and move the process into a proactive mode including implementing POMS.

**A–9.3. Description of the proactive obsolescence management programme**

*A–9.3.1. Scope of the programme*

The company's business platform, SAP, is used as a data source to proactively identify obsolescence. The data needed are component manufacturer, model, quality class. PM orders, BOM, component criticality and warehouse stock. These data are used to identify and to prioritize obsolescence issues. Selected data are sent to the POMS database in a secure way. There is a need for essential understanding by the management and organization of why the obsolescence programme is necessary and understanding the benefits of a proactive obsolescence programme.

Some manufactures and suppliers have a clear and proactive way of informing about upcoming obsolescence and the end of life of products. Sometimes there is no information available but even in those cases there could be an alternate product available. The industry tool POMS sometimes suggest a replacement from the OEM, or an alternative item used by another NPP.

*A–9.3.2. Responsibilities*

An obsolescence team was formed with an obsolescence manager assigned to establish and implement an obsolescence programme. The maintenance organization is responsible for the management of obsolescence and obsolescence engineers are also organized in the maintenance organization. Three obsolescence engineers (electrical, mechanical and I&C) are in the obsolescence team developing solutions. In case of need for technical assistants the issue can be handed over to the engineering department, for example when a design change is needed. The scope of prioritized obsolescence solutions is presented in a one year ahead report and approved by senior management. There is no obsolescence committee, instead the existing organization and working processes handle obsolescence issues. For example, when there is a need for extra budget or to develop a design change solution or a reverse engineering solution the coordination is done by the obsolescence manager, material planning and obsolescence engineers.

### A–9.3.3. *Prioritization process*

The existing data in the plant register set the scope and limitation for automatized prioritization in POMS. The following data are used:

— Safety classification;
— Criticality analysis;
— Seismic qualification;
— Quality class;
— Stock level;
— Installed base.

If there is a lack of spare parts or components that will affect the maintenance plans or safety stock level the priority will be high, and the obsolescence issue will be on the obsolescence team top list and included in the yearly planning. Basically, there are only two groups of prioritization: important or not important.

The template for prioritization is documented and approved by the management and is automated in POMS. However, this is only one source of information, and the final prioritization is done manually with more aspects included. The total scope of obsolescence solutions is compiled with input from the automated prioritization from POMS, system health and actual need from the maintenance organization.

### A–9.3.4. *Implementation of solutions*

Design change solutions are identified by the obsolescence team and when no other solution path is possible. In those cases, the development of the obsolescence solution is handed over to design change management. Depending on the complexity, the normal process is used to handle reverse engineering (often the design change process). Rebuild and repair are usually not considered that complex and are sometimes performed with a quality plan governing the work.

When implementing a proactive obsolescence process the lack of an IEE process became obvious. The IEE verification was in some way a part of other processes normally carried out when the equipment was installed in the plant. An IEE process was developed to support verifying equivalency and thereby made it possible to procure and store the new item on shelf. The IEE tool is essential to the obsolescence process and enables developing solutions proactively.

A process to support buying from the surplus market was also developed to support the obsolescence team. This process is mainly focused on identification and verification of the item being bought. The process is based on a graded approach depending on the complexity of the item being procured. Buying from the surplus market was not possible without this new routine.

### A–9.3.5. *Evaluation along the nine attributes*

TOP401 generally gives good guidance to form an obsolescence programme. Ringhals evaluated its obsolescence process along the nine attributes of TOP401 and determined the guidance in TOP401 was being met, with the following comments.

— Attribute 2A, Organization: It is possible to adapt the obsolescence process to the existing organization and it is not necessary to form an obsolescence committee if there are already forums in place that fit into the need of the obsolescence process. The resources at the plant are limited and prioritization does not have to depend on the root cause of the problem but rather the impact on the plant if no solution is implemented. The plant health needs to be prioritized and not depend on whether there is an obsolescence issue. There could be other needs for replacement, for example due to ageing or high failure rate. This can be a risk when using a separate committee to create a fast track for solving obsolescence issues.

— Attribute 8: INUOG needs to be added and offers a platform for obsolescence management with a focus outside the United States of America. INUOG and NUOG are partnering to avoid duplicate work and share experiences.

*A–9.3.6.  Data management in the programme*

No response provided.

*A–9.3.7.  Use of IT tools in implementation*

POMS is used as an IT tool for obsolescence management including KPIs (e.g. percentage obsolete items, industry benchmarks, solution progress and prioritization). The plant register and SAP, the business system, is also used together with POMS. External OEM databases with equipment in stock along with databases from third party engineering companies specializing in obsolescence solutions and replacement are also used. IT works well but is sometimes limited due to the data quality in the plant register.

*A–9.3.8.  Evaluation and maintenance of the programme*

The number of urgent issues is somewhat lower after a few years of obsolescence management and more focus is now on proactive work. No loss of production or scram has been caused by obsolescence as a root cause or contributing factor since the programme was in place.

Operating experience is fed back via INUOG and NUOG. Benchmarking of KPIs is done via INUOG and NUOG as well as a SALTO review of the process.

**A–9.4. Involvement of the regulatory body**

The obsolescence programme is well received by the regulatory body.

**A–9.5. Training**

The main source of training and understanding of obsolescence is through INUOG and NUOG. Those organizations work as a platform for discussion and development of processes, solutions, and tools to mitigate the effects of obsolescence. Operating personnel are encouraged to participate in these forums and along with internal training understand the obsolescence process.

**A–9.6. Operating experience with technological obsolescence**

International collaboration and sharing of knowledge and solutions through INUOG has saved a lot of time and made solutions possible (e.g. finding alternative parts and equivalent equipment).

# A–10. RESPONSE FROM SWEDEN (OKG)

## A–10.1. Description of the plant(s)

OKG operates Oskarshamn NPP in Sweden. The site consists of three BWR units, two of which are permanently shut down.

## A–10.2. Experience in development of an obsolescence management programme

In 2015, OKG decided to introduce a preventative programme for obsolescence. Prior to that, all obsolescence work had been reactive. Since 2017, OKG has a licence agreement to use POMS. When the agreement was signed, OKG sent all its 45,000 spare parts in the plant register (ODU) to POMS.

## A–10.3. Description of the proactive obsolescence management programme

### A–10.3.1. Scope of the programme

OKG uses ODU, an acronym for OKG′s operation and maintenance system, as a tool for obsolescence. ODU is the normal working tool of all maintenance engineers for preventive and reactive maintenance, which makes it easy for them to obtain up to date information on the obsolescence status of a spare part. Obsolescence has its own start page in ODU.

POMS makes frequent searches with suppliers and original manufacturers regarding which spare parts can be replaced or if a part is obsolete. The responses (obsolete, non-obsolete, additional data required or no response) are frequently entered into ODU via POMS. ODU is also continuously updated with new knowledge regarding obsolescence from, for example, maintenance engineers and spare parts engineers. It means that ODU always has updated information on the obsolescence status of all spare parts.

OKG uses POMS for information exchange. If there are solutions for obsolete components in POMS (additional solutions, refurbish, available at another site) it is entered into ODU. If OKG has a solution for an obsolete component, it is returned to POMS. A spreadsheet is used for communication between ODU and POMS.

An important part is to update the item information on components that POMS, manufacturers or suppliers cannot identify. It is clear from POMS which articles need to be updated. For priority articles, maintenance engineers receive this assignment via steering committee meetings.

To get the most out of the obsolescence programme and the industry information tool, data improvement work has to be done to make sure that the data meet the demands from the obsolescence team. Sometimes plant walkdowns are necessary and in some cases onsite information can be used to correct the data. The present data could have been used despite missing information, spelling errors and incorrect manufacturer data (e.g. the correct branding is not used because the contractor who was responsible for design used their own name instead of the OEM).

### A–10.3.2. Responsibilities

The steering committee for obsolescence is a cross-functional AMP group led by the obsolescence manager. The steering committee coordinates the work on obsolescence between the relevant functions at OKG and is represented by the obsolescence manager in the coordination group for ageing management. Participants in the steering committee are the obsolescence manager, five maintenance managers from Unit 3: (Mechanical (3), Electrical, I&C) and the manager for ODU application and spare parts. The steering committee has scheduled meetings every month. If the cross-functional knowledge is insufficient, experts are called to the meetings.

The spare part engineers are the contact for purchases. They write the purchase request, with the technical requirements that apply to an article. The maintenance engineer writes a technical request in his assignment. The technical department uses the technical request to prepare their design work.

### A–10.3.3. Prioritization process

ODU grades and prioritizes all articles automatically in real time. The articles are graded if they are linked to sub-objects in safety class S1–S3 or availability class T1. The grading includes also if the spare part is obsolete, requires additional master data, or no response in POMS. Examples of grading are:

— Articles S1–S3: with repair criteria, found obsolete;
— Articles S1–S3: with repair criteria, POMS needs additional master data;
— Articles S1–S3: with repair criteria, no response from POMS.

The spare parts are then prioritized as 'red', 'yellow' or 'green'. Prioritization is made with respect to inventory balance and stock withdrawals over the past 10 years. Examples of prioritization are:

Priority RED

Stock balance = 0 and stock withdrawal last 10 years > 0

Priority YELLOW (1)

Stock balance = 0 and stock withdrawal last 10 years = 0

Priority YELLOW (2)

Stock balance> 0 and stock withdrawals for last 10 years > current stock balance

Priority GREEN

Stock balance> 0 and stock withdrawal last 10 years <= current stock balance

An ongoing top list is produced by the steering committee. The document is based on grading, prioritization, risk analysis and experience. Follow-up and decisions on new assignments for the top list is a recurring point at the obsolescence group's monthly meetings. The maintenance managers in the steering committee take on new assignments at the meeting and delegate them to their engineers. The obsolescence manager is a support for the engineers with possible options for dealing with obsolete spare parts.

For the prioritized spare parts where the overall competence in the steering committee cannot assess the spares parts ranking on the top list, a risk analysis from the operating department is required.

### A–10.3.4. Implementation of solutions

To implement an obsolescence solution, it always starts with the assigned maintenance engineer writing a technical question. Then it will be handled in accordance with the procedures for a design change.

*A–10.3.5. Evaluation along the nine attributes*

Attribute 1 – Overall Understanding:

— Does the programme and employees have the prerequisite for understanding when an obsolete article is considered obsolete? Yes, 2016–14638 - Proactive programme for obsolescence, describes the process of obsolescence. Obsolete articles are updated in ODU.
— Does the programme and employees have the preconditions for understanding whether an article in storage can be obsolete? Yes, 2016–14638 - Proactive programme for obsolescence describes the process of obsolescence and inventory balance's impact on article priority. Obsolete articles are updated in ODU.
— Does the programme and employees have the prerequisite for understanding if an article on which the supplier changes the part number is obsolete? Yes, 2016–14638 - Proactive programme for obsolescence describes the process of obsolescence. Obsolete articles are updated in ODU. If the supplier changes the part number, it is an entirely new item on OKG, which needs to be approved.
— Does the programme and employees have the prerequisite for understanding if an article that the supplier cannot provide sufficient documentation on, which is already in the facility, is obsolete? Yes, 2016–14638 - Proactive programme for obsolescence describes the process of obsolescence. Obsolete articles are updated in ODU.
— Does the programme and employees have the prerequisite for understanding that an article is obsolete, even when the supplier has a developed model with better specifications? Yes, 2016–14638 - Proactive programme for obsolescence describes the process of obsolescence. Obsolete articles are updated in ODU. If the supplier changes the part number or develops a new model, it is a completely new article on OKG, which needs to be approved.

Attribute 2 – Preventive Measures:

— Does the programme provide the employee with the conditions for identifying obsolete articles? Yes, 2016–14638 - Proactive programme for obsolescence describes the process of obsolescence. Obsolete articles are updated in ODU.
— Does the programme provide the employee with the preconditions for prioritizing the identified obsolete articles? Yes, 2016–14638 - Proactive programme for obsolescence, describes the priority model for obsolete articles. In ODU, all articles are graded and prioritized.
— Does the programme provide the employee with the means to solve the problems with the priority articles? Yes, 2016–14638 - Proactive programme for obsolescence, 2016–07101 - Agreement with POMS, OKG's project model (PUP), 2005–09919 - Maintenance at OKG, and membership in INUOG.

Attribute 3 – Detection of Obsolescence:

— Is there an effective process for detecting obsolescence before SSC failures occur? Yes, 2016–14638 - Proactive programme for obsolescence. Obsolete articles are tagged in ODU. 2016–07101 - Agreement with POMS. Maintenance engineers experience from other power plants. Membership in INUOG.

Attribute 4 – Monitoring and Trending Obsolescence:

— Are there condition indicators and parameters that are monitored? Yes, 2016–07101 - Agreement with POMS.

— Is there information to be collected to facilitate the assessment of obsolete SSC? Yes, 2016–07101 - Agreement with POMS.
— Are there any measurement methods on the effectiveness of the obsolescence program? Yes, KPIs are developed.

Attribute 5 – Mitigation Efforts:

— Is there an action programme with action plans, administrators, and completion time, for priority obsolete items? Yes, a top list is created and managed by the steering committee at monthly meetings.

Attribute 6 – Acceptance Criteria:

— Are there any acceptance criteria that determine the need for corrective action for obsolete articles? Yes, 2016–14638 - Proactive programme for obsolescence. Grading, prioritization, and acceptance criteria for all articles are described.

Attribute 7 – Corrective Action:

— Is there an action programme with action plans, administrators, and completion time, for priority obsolete items? Yes, a top list is created and managed by the steering committee at monthly meetings.

Attribute 8 – Experience Feedback:

— Are there processes that ensure quick feedback on experiences and knowledge about obscene articles? Yes, ODU is updated immediately to indicate articles being obsolete, whether it is experience from maintenance engineer, spare part engineers or results from POMS.

Attribute 9 – Quality Assurance:

— Are there administrative controls that document the implementation of obsolescence and the measures taken? Yes, obsolescence is an AMP group, which reports its work on coordination meetings for ageing, 2015–25332 - The Ageing Management Coordination Group.
— Are there indicators to facilitate evaluation and improvement of obsolescence management? Yes, KPIs are developed.
— Is there a verification process to ensure that preventive measures are adequate and appropriate and that all corrective actions have been completed and are effective? Yes, KPIs are developed.
— Are there documentation and practices to follow? Yes, 2016–14638 - Proactive programme for obsolescence. 2015–25332 - The Ageing Management Coordination Group.

*A–10.3.6. Data management in the programme*

OKG uses ODU as a tool for obsolescence. POMS is used for information exchange. An Excel sheet is used for communication between ODU and POMS.

*A–10.3.7. Use of IT tools in implementation*

ODU is the only IT tool and database used.

Experience with the use of IT tools: Only the IT tools that are known and used by all maintenance engineers are used at the NPP. New programmes or databases are not introduced.

*A–10.3.8.  Evaluation and maintenance of the programme*

Very good experience that obsolescence was implemented in ODU, where all maintenance engineers are used to working. Operational experience is utilized by documenting everything in ODU. Efficiency is maintained by KPIs with monthly follow-up on status and progress in the work on obsolete articles.

## A–10.4. Involvement of the regulatory body

The regulatory body was involved when an operational obsolescence programme was a prerequisite for continued operation of the NPP.

## A–10.5. Training

The obsolescence manager has held information meetings about obsolescence and how the start page is structured in ODU, for each maintenance group. The obsolescence manager has continuous follow-up meetings with engineers who are given assignments. This is to support and show interest in results.

## A–10.6. Operating experience with technological obsolescence

There is an agreement with an important supplier to ensure competence and access to spare parts for 40 years. This involves preventive work with obsolescence, the development of replacement components and rapid feedback on the replacement of obsolete components that will involve redesign of the machines.

## A–11.  RESPONSE FROM SWEDEN (FORSMARK)

## A–11.1. Description of the plant(s)

Vattenfall operates Forsmark (FKA) NPP, which has three BWR units and is located on the east coast of Sweden.

## A–11.2. Experience in development of an obsolescence management programme

After participating in an IAEA Technical Meeting in 2017, FKA started to develop a proactive obsolescence management programme in 2018. At the same time, FKA started to participate in INUOG activities, which gave much input in the development of the programme. FKA also got great support from Ringhals, who had been developing their proactive obsolescence programme since 2015. FKA and Ringhals now share more or less the same obsolescence management organization structure and procedures.

In 2018, FKA also procured access to the POMS database to aid in detecting, prioritizing, and solving obsolescence at the site.

The IGALL TOP401 was used as a reference when forming the obsolescence programme.

The obsolescence programme was not fully implemented as of April 2020. A challenge has been to fit the programme to the existing organizational structure. The interface to other parts of the organization is not completely defined. Several initiatives to solve obsolescence were started before the obsolescence programme existed and have not yet been incorporated into the obsolescence programme, which has made it hard to get a full overview of the obsolescence situation at the plant. It has sometimes taken a long time to get the obsolescence status of models from vendors and manufacturers.

## A–11.3. Description of the proactive obsolescence management programme

### A–11.3.1. Scope of the programme

Since FKA and Ringhals belong to the same organization, the same asset management system, SAP, is used at both NPPs. Selected plant and maintenance data (e.g. component criticality, component data, and stock data) are sent to the POMS database and obsolescence identification, prioritization, and solution documentation is done through the POMS web interface.

With the right information, the OEM is often able to provide suggestions for replacement parts. Merging of OEMs makes it more difficult to get information about the original model due to a change of product lines or a lack of interest from the new manufacturer to look into old documents to find information to be able to suggest a replacement part.

To identify missing information, photos of equipment in the plant taken by maintenance personnel has been used. The photos can show essential information that makes it possible to identify the model. FKA also has photos of spare parts in the warehouse. These can be helpful in finding a model number.

### A–11.3.2. Responsibilities

An obsolescence manager with a team of two obsolescence engineers and material planners manages the obsolescence at FKA. The group already existed within the maintenance department with a material planning function but was extended to also manage obsolescence.

The obsolescence engineers identify and prioritize obsolescence, and in some cases find a solution. They also work with data cleanup in the system. If technical assistance is needed, the Engineering Department gets involved. A separate group within the Maintenance Department manages more complex solution actions like item equivalency evaluations, reengineering, and rebuild and repair. The obsolescence manager, with the obsolescence engineers, coordinate with the whole supply chain.

A top ten list of obsolete components is presented to an obsolescence committee (the same committee that approves projects at the site, consisting of plant senior management from engineering, operations, and maintenance) twice a year, and a decision is made on what to prioritize.

### A–11.3.3. Prioritization process

Obsolescence is prioritised by:

— Safety classification;
— Install count;
— Environmental qualification;
— Quality class;
— Seismic class;

— Expected future demand;
— Historical work order count;
— Work order priority;
— Stock level;
— Inventory status (stock, do not stock).

In addition, component criticality and SPV is taken into account. However, component criticality and SPV identification was done recently and has not yet been added to the asset management system. SPV identification is also currently only at a functional group level.

The prioritisation is automatically done in POMS by the OVR algorithm, which sets points and weights to the criteria and calculates a total OVR score for each component. The OVR algorithm can be adjusted by the management. Changes to the OVR algorithm are documented. POMS groups the components into four priority categories, but at the moment, FKA only consider the top category (red) as important, and the other categories as not important. A final priority, which may differ from the automatic OVR priority, is set by the obsolescence team and the obsolescence committee.

*A–11.3.4. Implementation of solutions*

Since before the obsolescence programme was initiated, an organization existed that was responsible for design change and equivalency evaluations. FKA was also in parallel evaluating a process for reverse engineering and was negotiating with manufacturers for a special manufacturing run. There have also been cases of procurement from surplus markets and cannibalization. These procedures need to be documented so that they can be used as a solution strategy for obsolescence.

*A–11.3.5. Evaluation along the nine attributes*

The obsolescence programme at FKA has been organized around the nine attributes in the IGALL TOP401, and the programme has been evaluated against those attributes. Some key observations are:

— Attribute 2A, Organization: The existing organization, with the addition of the obsolescence manager and obsolescence engineer roles, has been adapted to the obsolescence programme. Even though obsolescence was managed mostly reactively before, the functions to solve obsolescence issues were present in the organization.

— Attribute 2B, Methodology: A lot of effort has been put into data cleanup in the plant register and identifying unknown components. The identification of obsolete items is more organized than before and everyone at the plant is encouraged to notify the obsolescence team of possible obsolete components. An obsolescence email address has been created for this purpose. Identification of obsolescence is also done by vendor contact. Another source to detect obsolescence is the POMS database and monthly call meetings with INUOG.

— Attribute 3: The source of information about obsolescence mainly comes from vendor contacts. The procurement department will sometimes report signals of possible future obsolescence, but there is no structured method to make this information widely known in the organization. The obsolescence team attends monthly call meetings with INUOG, where information about potential obsolescence issues is shared among the members.

— Attribute 4: An assessment of the obsolescence programme against the TOP401 attributes was made early in the programme development to identify areas that need further development.

KPIs (e.g. unsolved obsolete components, unsolved SPVs, critical components requiring data cleanup) have been developed and are reported to management.

— Attribute 5: Equivalent replacement has been used at FKA many times, and the method is documented. Depending on the complexity of the component to be replaced, it can be very expensive.

Buying from the surplus market can be complicated due to unique requirements on the components from FKA. There is no documented procedure for buying from the surplus market, and each case is handled individually.

A successful pilot project using reverse engineering in collaboration with an experienced supplier has been completed at FKA. That resulted in a documented procedure that will be used in the future.

FKA has a well-equipped workshop that can repair and manufacture new parts. This is often a more cost effective solution than an equivalent replacement or a special manufacturing run.

— Attributes 6 and 7: Since the programme has only been in operation for less than a year, the use of acceptance criteria and corrective actions has not been tested.

— Attribute 8: FKA actively participate in INUOG workshops, seminars and monthly call meetings to share lessons learned. Since INUOG and the US-based NUOG have a close collaboration, lessons learned from a large part of the industry are shared among the members.

*A–11.3.6. Data management in the programme*

All data about obsolescence and potential solutions are stored in the POMS database. As of April 2020, no feedback exists to the FKA plant asset management system Fenix/SAP; however, such feedback is planned for the future.

*A–11.3.7. Use of IT tools in implementation*

The POMS database with web interface is used to manage obsolescence. The built-in functions to produce reports with KPIs for senior management is also used. In parallel, the plant register and maintenance system Fenix is used, but there is no feedback from POMS to Fenix. Such feedback would be of great value to spread information about obsolescence status in the organization and is planned for the future. Data cleanup is done in Fenix.

*A–11.3.8. Evaluation and maintenance of the programme*

The obsolescence programme has been operational for about a year, and the focus has mainly been on data cleanup. Thus, there is limited operational experience in the programme. No assessment, besides the initial TOP401 assessment, has been made.

Operational feedback is done through INUOG.

## A–11.4. Involvement of the regulatory body

The regulatory body reviewed the initial programme while the plant was preparing for an IAEA pre-SALTO mission. Some parts of the programme had not yet been implemented at that time.

## A–11.5. Training

Active participation in the organizations INUOG and NUOG is encouraged by the management. Most training of the staff is through seminars and meetings organised by INUOG. This has worked very well, since there are many experienced members represented in INUOG.

## A–11.6. Operating experience with technological obsolescence

FKA has secured the availability of spare parts for a certain model for the plant lifetime by contracting with a manufacturer. This was made before the current obsolescence programme was established.


## A–12. RESPONSE FROM SWITZERLAND

### A–12.1. Description of the plant(s)

Information was provided by two of the Swiss NPPs: Leibstadt (denoted as KKL), and Gosgen (denoted as KKG). KKL is a single unit BWR site and KKG is a single unit PWR site.

### A–12.2. Experience in development of an obsolescence management programme

Obsolescence management has historically grown and been implemented in different ways in the different NPPs of Switzerland. Although not implemented under the specific term obsolescence management, there are measures in place to address the issue. Therefore, this section contains information from two of the three operational NPPs of Switzerland to provide insight into the applied solutions in both.

At KKG, a structured process for obsolescence management has been introduced (although it is not referred to as this), formed out of existing processes. The totality of these processes works in the same way as obsolescence management. The basis in the department of electrical engineering and maintenance is the so-called living PSU. Here, each responsible person summarizes the past year in a systematic way, assesses their systems, and evaluates their condition with regard to ageing, spare parts, etc. This structured information is used as an input to the change management process, which is the process within KKG where technical changes, replacements, or refurbishments are planned and prioritized. At KKG, the living PSU started in 2017 as part of the periodic safety review in 2018. In 2019 and 2020, as well as in future years, these data will be updated periodically and systematically. The experience is very positive so far.

KKL is currently working on the topic of obsolescence management. A review was recently performed by EPRI. It confirmed that the implemented approach seems to manage obsolescence issues adequately. No case was detected where obsolescence was primarily responsible for impacting power generation or safety. The current approach is mainly based on very experienced, knowledgeable, and dedicated staff as well as on proven processes, procedures, and practices. Cooperation between the various organizations seem to be effective and good information systems are in place. However, it was also identified that the current approach to obsolescence might rather be tactical and that today no formal, documented programme or process specifically dedicated to managing obsolescence is in place. The potential for improvement is currently under consideration.

The implemented approach has proven effective and satisfactory to date. As the plant is ageing, and original parts or manufactures become more and more unavailable, the topic of proactive obsolescence management seem to gain importance. As explained by KKG, and also KKL, the living PSU plays an important role in the management of obsolescence.

## A–12.3. Description of the proactive obsolescence management programme

### A–12.3.1. Scope of the programme

At KKG, the basis in the electrical engineering department is the periodically performed living PSU. In other departments the source of data may be different.

At KKL the living PSU also plays an important role. In addition, further processes are in place to collect relevant data. This relates for example to the assessment of the conditions of installed equipment, the monitoring of parts availability and contact with suppliers (e.g. regarding availability of products and services). Data are maintained in different systems. EPRI highlighted in their review the good availability of relevant data. Document reviews, walkdowns, and supplier meetings are regularly performed. The implemented AMP is also a significant element and contributor.

### A–12.3.2. Responsibilities

At KKG, each item of equipment has an owner who is responsible for all technical aspects such as correct operation, periodic functional tests, ageing management, planning/proposal for systematic replacement etc. For mechanical equipment, the responsibility is in the mechanical engineering/maintenance department, for electrical equipment the responsibility is in the electrical engineering/maintenance department.

At KKL, responsibilities are distributed across the organizations involved, such as engineering, maintenance and purchasing. Also, a system is in place in which responsibilities are defined for systems and components. The responsible persons are important drivers and key players in the process of obsolescence management. KKL has also implemented a systematic process to involve plant senior management. One example is the 'BEDAM' process, in which (strategic) decisions are made regarding the management of the operational lifetime of systems and components. Experts and working groups are also involved in this topic (e.g. the cable group, consisting of experts in the lifetime management of the installed cabling). There is a close cooperation in place between purchasing and engineering/maintenance; both maintain close contacts with key suppliers.

### A–12.3.3. Prioritization process

At KKG, the proposals for replacements and changes of items, including possible solutions, are produced in the technical departments within the overall Change Management Process. It is a structured process consisting of milestones and gates. The final decision on prioritization as well as on the decision to implement a change or not is with the Project Portfolio Board, which consists of all relevant department managers (Electrical, Mechanical, Supervision, Operation, Long term Operation, Safety, Supply Management, Finance, etc.) as well as the station director of KKG.

KKL has also implemented a systematic process. Proposals are typically elaborated by the technical departments, under the lead of the department management. Further consolidation of projects is done at plant level by the plant management, where the department managers are represented.

### A–12.3.4. Implementation of solutions

At KKG, the technical solution, including weighting of the possible solutions, is worked out by specialists in the technical departments, often in cooperation across the various departments and in cooperation with potential suppliers.

At KKL, a similar process is in place, in which appropriate options for solutions are evaluated during a concept phase and the solution chosen is then implemented via a project (e.g. project for gradual replacement of safety control system).

*A–12.3.5.  Evaluation along the nine attributes*

No response provided.

*A–12.3.6.  Data management in the programme*

At KKL, information exchange between the different organizations involved is based on internal procedures and good practices. Furthermore, KKL contributes to, and cooperates with, different industrial organizations, such as the Equipment Qualification Databank and EPRI. A close collaboration is implemented between the Swiss NPP operating organizations.

*A–12.3.7.  Use of IT tools in implementation*

The KKG change management process is supported by a tool. The tool is included in the ERP of KKG (the enterprise resource planning tool is currently IPIS).

The KKL process is supported by various tools. The most important tools IT database tools in this regard are the IPIS and SAP ERBs.

*A–12.3.8.  Evaluation and maintenance of the programme*

KKG has a management system, which is ISO 9001/2015 certified. There are supplier audits, internal audits and self-assessments defined. All processes are subject to the management system, including the change management process.

As mentioned before, at KKL obsolescence management is not implemented as a single, dedicated process with exactly that name but is a result of various processes as part of the overall Total Quality Management System. The management system is, among other things, ISO 9001/2015 certified and maintained accordingly (including self-assessments, internal and external audits, supplier audits, etc.). Effectiveness of the implemented obsolescence management process was recently confirmed by an EPRI review.

## A–12.4. Involvement of the regulatory body

KKG is accountable to the Swiss regulatory body for the long term strategy for systematic replacement of systems. The new LTO department was set up, among other reasons, for this purpose.

KKL is also accountable to the regulatory body, which has implemented clear guidelines (although not directly under the name obsolescence management) and performs associated reviews and inspections. The regulatory body is involved in the approval of replacement solutions and new projects and installations. Furthermore, the regulatory body is active in requesting and reviewing KKL strategies for LTO.

## A–12.5. Training

The systematic development of the change management process at KKG was designed over a five year period. Numerous process trainings were carried out by the process owner. The technical knowledge of the engineers is fundamental and therefore systematic technical training is also carried out.

Systematic training of personnel is also an important task at KKL, which is given a high priority. This involves not only the training of processes and procedures but also the systematic development of (technical) knowledge, skills, etc. A process is implemented for this, and a dedicated function is allocated in the electrical department. The process is regularly reviewed by the regulatory body.

## A–12.6. Operating experience with technological obsolescence

A practical example of this at KKL is the ongoing project for the modular replacement of the installed safety control system. For this, an intensive concept study was performed, and different possibilities were evaluated. A systematic decision process, involving senior management and experts, provided the main framework for a project. This project has begun and will be implemented over the coming years.

## A–13. RESPONSE FROM THE UNITED KINGDOM (EDF-UK)

### A–13.1. Description of the plant(s)

EDF operates a fleet of NPPs in the United Kingdom. This includes eight Advanced Gas Reactors and one PWR. Construction of two PWRs are ongoing with a further two having received permission for construction to progress.

### A–13.2. Experience in development of an obsolescence management programme

EDF restarted an obsolescence management programme in 2015, after identifying an area for improvement for obsolescence management. The obsolescence management process has been revised multiple times to take into account improvements in prioritization, alignment with organisational changes, external operating experience and from performance improvement activities. In 2016 the wider roll out of the POMS obsolescence software tool commenced. POMS is a web-based service that provides the ability to view up to date obsolescence issues by component, system, criticality, and stocking level. The tools are configured to allow both local and fleet level report review of obsolescence issues and system health status to be identified and reported.

In 2018, EDF started arrangements for extended operation of the PWR and planned LTO until 2055. To support continuous improvement, EDF has developed a data cleanup targeted approach for high criticality equipment (following AP 913 Equipment reliability) and SPV equipment. A focused project has been undertaken enhance the manufacturer/model data quality significantly. Design document reviews and walkdowns are performed to resolve missing information.

Challenges experienced through development of the obsolescence programme include:

— Reorganisations resulting in changes in roles and responsibilities within the obsolescence programme governance;
— Prioritization programme arrangements due to changing station lifetimes;

— Resource availability to support proactive obsolescence resolutions.

## A–13.3. Description of the proactive obsolescence management programme

### A–13.3.1. Scope of the Programme

EDF collects data from the plant asset management system and POMS to support proactive obsolescence identification. The following processes are involved:

— Previous procurement: If the part was purchased within the past three years, it is noted as not obsolete in the obsolescence management programme.
— Vendor substitution through procurement process allows for the OEMs to identify replacement parts.
— POMS industry replacement solutions: This includes the OEM, alternate manufacturer, and solutions progressed by other NPPs.
— POMS spare parts availability status: As part of POMS confirming obsolescence of a part, they also ask the OEM if they still support spare parts.

### A–13.3.2. Responsibilities

The EDF obsolescence management programme is directed by the engineering organization. Within the integrated arrangements for obsolescence, there are obsolescence process leads at each operating location that coordinate, engage, and interact with the rest of the organization. The leads hold obsolescence screening and solution progress meetings to proactively manage top obsolescence issues. The obsolescence management programme assists with identifying obsolescence issues on the work plan, which is then resolved through the permit request process.

The obsolescence fleet working group is an internal team of leads that has direct oversight of the obsolescence process and performance. Obsolescence key performance indicators are presented to the station Plant Health Committee and central oversight committee periodically (annually).

### A–13.3.3. Prioritization process

POMS is used to provide a numerical value for prioritising new obsolescence occurrences and local or fleet working groups then add intelligence to the review and sanction a priority bi-monthly.

### A–13.3.4. Implementation of solutions

EDF reviews solution options following a recommend hierarchy for solution pathways for resolving obsolescence based on cost and complexity of the required solution. Due to the recent life extension, replacing an obsolete part with a commercially available part is preferred and prioritized over a repair strategy. For the replacement strategy, the replacement part is evaluated through an item equivalency screening process to determine if a short or long form item equivalency can be performed or if a design change is needed.

### A–13.3.5. Evaluation along the nine attributes

EDF has taken learning from both other utilities and the IAEA TOP401 document and incorporated the relevant content into the obsolescence management governance. Notable content incorporated into the governance was:

— Structured identification of occurrences;

— Metrics and monitoring;
— Obsolescence process documents and hierarchy of process steps and connections;
— Process structure and reporting milestones;
— Risks and opportunities from various solution pathways;
— Identification of obsolete components in EDF master asset database (including work completed on obsolete item);
— Early warning indicators of obsolete parts.

*A–13.3.6. Data management in the programme*

No response provided.

*A–13.3.7. Use of IT tools in implementation*

EDF has been using the Obsolescence Manager module in POMS to identify obsolescence risks and document and track obsolescence solutions through the mitigation stages. EDF has developed internal software to also monitor programme progress, including identification and prioritization. All data are stored on company server infrastructure allowing integration with other applications and organizations.

*A–13.3.8. Evaluation and maintenance of the programme*

EDF has used KPIs based on industry best standards to monitor and trend performance. KPIs include:

— Overall availability status of components (Obsolete, Not Obsolete, Unknown or Unusable);
— Completed Obsolescence Solutions;
— Obsolescence Solutions Ready for Execution.

## A–13.4. Involvement of the regulatory body

No response provided.

## A–13.5. Training

EDF has developed a number training items for the obsolescence programme and implemented them as required to meet the local NPP needs.

## A–13.6. Operating experience with technological obsolescence

Solution examples provided in main body of document. No additional information.

# ABBREVIATIONS

| | |
|---|---|
| AMP | ageing management programme |
| BOM | Bill of Materials |
| BWR | boiling water reactor |
| EPRI | Electric Power Research Institute |
| ERP | enterprise resource planning |
| IEE | item equivalency evaluation |
| IGALL | International Generic Ageing Lessons Learned |
| INUOG | International Nuclear Utility Obsolescence Group |
| I&C | instrumentation and control |
| KPI | key performance indicator |
| LTO | long term operation |
| NUOG | Nuclear Utility Obsolescence Group |
| NPP | nuclear power plant |
| OEM | original equipment manufacturer |
| OVR | obsolescence value ranking |
| PLC | programmable logic controller |
| PWR | pressurized water reactor |
| POMS | Proactive Obsolescence Management System |
| RTD | resistance temperature detector |
| SALTO | Safety Aspects of Long Term Operation |
| SPV | single point vulnerability |
| SSCs | structures, systems and components |

# CONTACT IAEA PUBLISHING

Feedback on IAEA publications may be given via the on-line form available at:
www.iaea.org/publications/feedback

This form may also be used to report safety issues or environmental queries concerning IAEA publications.

Alternatively, contact IAEA Publishing:

Publishing Section
International Atomic Energy Agency
Vienna International Centre, PO Box 100, 1400 Vienna, Austria
Telephone: +43 1 2600 22529 or 22530
Email: sales.publications@iaea.org
www.iaea.org/publications

Priced and unpriced IAEA publications may be ordered directly from the IAEA.

## ORDERING LOCALLY

Priced IAEA publications may be purchased from regional distributors and from major local booksellers.

25-06198E

Printed and bound by CPI Group (UK) Ltd, Croydon, CR0 4YY

06/07/2026

02160600-0005